T0207448

Graphs, Colourings and the Four-Colour Theorem

GRAPHS, COLOURINGS AND THE FOUR-COLOUR THEOREM

Robert A. Wilson

The University of Birmingham

OXFORD
UNIVERSITY PRESS

This book has been printed digitally and produced in a standard specification
in order to ensure its continuing availability

OXFORD
UNIVERSITY PRESS

Great Clarendon Street, Oxford OX2 6DP

Oxford University Press is a department of the University of Oxford.
It furthers the University's objective of excellence in research, scholarship,
and education by publishing worldwide in

Oxford New York

Auckland   Cape Town   Dar es Salaam   Hong Kong   Karachi
Kuala Lumpur   Madrid   Melbourne   Mexico City   Nairobi
New Delhi   Shanghai   Taipei   Toronto
With offices in
Argentina   Austria   Brazil   Chile   Czech Republic   France   Greece
Guatemala   Hungary   Italy   Japan   South Korea   Poland   Portugal
Singapore   Switzerland   Thailand   Turkey   Ukraine   Vietnam

ISBN 0-19-851062-4

Preface

This book arose out of a third-year module in graph theory given at the University of Birmingham over the three years 1996–9, and again in 2001. This module was designed to be accessible to a large number of students (the prerequisites are minimal), but still to present some challenging material.

The course centres around the famous 'four-colour conjecture', that every map can be coloured with four colours, subject to the usual convention that no two adjacent countries may be coloured the same. From its first appearance in mathematical folklore in the 1850s, until its eventual solution in the 1970s, this apparently simple problem has frustrated generations of mathematicians, both professional and amateur.

The book begins with a discussion of the early approaches of Kempe and Tait in the 1870s and 1880s, before revealing the flaws in their arguments, and then describing some of the ways in which the methods were refined, the problems axiomatised, and the conjectures generalized. In the course of this, we present several of the finest gems of the subject: Heawood's bound for map-colouring on a surface with holes, Kuratowski's theorem characterising which graphs (or maps) can be drawn on a surface without holes, and Vizing's theorem on the minimum number of colours needed to colour the edges of a graph. The final part of the book aims to provide some insight into the methods which eventually cracked the four-colour problem.

Much of the material in this book was covered in a single course of about 20 lectures, although some extra material has been added for completeness, and to facilitate a personal selection of topics. If students have met graphs before, then Chapter 2 can be largely omitted. If the aim is to study the four-colour theorem itself in some depth, then Chapters 7 and 8 are somewhat tangential and can also be omitted. On the other hand, a more general graph theory course can be made by picking a somewhat broader mix of topics from all the chapters.

I am grateful to the many students who took my course for their help in removing errors and in stimulating me to better exposition. In particular I would like to thank Tamar Watts, Stuart Underwood, Kate Stowe, Clare Robinson, Richard Barraclough and Alan Barclay for their meticulous attention which uncovered a number of errors. Thanks go also to my colleagues, especially Chris Parker and Tony Gardiner, for their constructive criticism and helpful suggestions, as well as to the anonymous referees, who made many useful comments which I believe have led to significant improvements. Needless to say, I take the blame for all the errors which remain. Finally, thanks to Elizabeth Johnston, Ruth Walker, and everybody else at Oxford University Press who helped transform my lecture notes into this book.

Contents

Part I

Graphs, maps and the four-colour problem

1
Introduction

1.1 Preliminaries

In this book, we will study **graph theory** with particular reference to colouring problems. Perhaps the most famous graph theory problem is the **four-colour conjecture** (4CC), first stated by Guthrie in 1852, widely publicized by Cayley in 1878, but only solved in 1976 with computer assistance (so now we can call it the **four-colour theorem**, or 4CT). We will study this and related problems.

First, we state the four-colour problem. This can be expressed in a variety of different ways, and we start with a rather informal version, before giving a more mathematically precise version later on. A **map** consists of **countries** bounded by simple closed curves, and we say that two countries are **adjacent** if they have a common border which contains at least a segment of a curve, not just a finite collection of isolated points. We wish to colour the countries so that any two adjacent countries have different colours. The problem is, how many colours do you need in order to be able to colour all maps in this way? In order that the problem should have an answer, we need to make a few restrictions on what constitutes a map. For example, each country must be connected, or else there is no bound on the number of colours which might be required.

The outside region of the map is also considered a country, although this makes only a technical difference to anything. By considering the map in Fig. 1.1, we see that four colours are certainly necessary, as each of the four countries labelled A, B, C, D is adjacent to each of the other three.

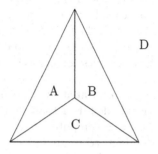

Fig. 1.1 A map which is not 3-colourable.

1.2 History of the four-colour problem

It is worth noting that, despite many assertions to the contrary, there is no evidence of this question being raised by map-makers before it was raised by mathematicians. After the first recorded statement of the problem/conjecture by Francis Guthrie in 1852 (communicated by his brother Frederick to Augustus De Morgan, and recounted in a letter from De Morgan to Hamilton on 23rd October), it remained in some obscurity until Cayley drew attention to the problem again in 1878, in the Proceedings of the London Mathematical Society. It was then that mathematicians realized the problem had not yet been solved, and some effort was put into its solution. At this point, it was generally thought that the problem was not seriously hard, and would soon be solved.

Kempe provided a clever argument in 1879 which purported to prove that four colours were sufficient. It took some 10 years before the error was detected, by Heawood, who then patched up the proof to show that five colours were sufficient. (Heawood continued publishing occasional papers on the four-colour problem until 1949!) The problem remained open, then, to determine whether four colours were enough, or whether there was some map which required five colours. It was only at this stage that mathematicians began to feel that perhaps the problem was harder than was at first thought.

Progress was slow. The two main ingredients in the eventual proof by Appel, Haken and Koch [4,5] in 1976 were 'reducibility' and 'discharging', which we explain in Chapter 9. The concept of reducibility was introduced formally in 1913 by Birkhoff [12], who managed to prove that many configurations were reducible. Essentially this means that the configuration can be reduced to a smaller case, which by induction we can assume to be 4-colourable.

The idea of discharging is due to Heesch, who really came quite close to proving the four-colour theorem, and should be given more credit than he normally is for his part. (Ore, whose very influential book [37] on the four-colour theorem was published in 1967, was apparently unaware of the work of Heesch.) Essentially, one uses a kind of conservation law to show that if a graph **globally** fails to satisfy the four-colour theorem then there is some **local** obstruction. Thus, 'discharging' produces a long list of these local obstructions, or 'unavoidable configurations' as they are called—that is, every graph (or at least every minimal counterexample to the four-colour theorem) must contain at least one of these configurations. Then each of these unavoidable configurations should be proved reducible. This then provides an inductive proof of the four-colour theorem.

The first proof used a very complicated discharging algorithm, devised by hand, which produced an unavoidable set of 1936 configurations, each of which was then proved irreducible, using a computer. It then turned out that 102 of these configurations were redundant, so the number required for the proof was just 1834. Later, this number was reduced still further, to 1482. A more recent simplification of the proof by Robertson, Sanders, Seymour and Thomas [42] used an unavoidable set of only 633 configurations.

More details of the history of the problem can be found in the books by Barnette [7], Fritsch and Fritsch [23] and Biggs, Lloyd and Wilson [10]. The latter includes the original papers of Cayley and Kempe, as well as large parts of Heawood's paper, and other relevant papers up to 1936.

Exercises

Exercise 1.1 Find a map which cannot be coloured with three colours, but does not have four mutually adjacent regions.

Exercise 1.2 (The empire problem) Consider the problem of colouring a map of empires, in such a way that all the countries in a given empire are coloured with the same colour. Construct a map of n empires which requires n colours. (Each empire may consist of as many countries as you like.)

2
Basic graph theory

2.1 Some definitions

Before we get onto the real subject matter of the book, we revise the basic definitions of graph theory. It is unfortunate that even the most basic of these definitions is not entirely standardized, to the extent that not everyone agrees even on the meaning of the word 'graph'. Thus, you need to be aware of possible differences in meaning when comparing with other sources.

Definition 2.1 *A **pseudograph** G consists of a set $V(G)$ of **vertices** and a set $E(G)$ of **edges**, such that each edge is **incident with** two (not necessarily distinct) vertices. The edge is then said to **join** these two vertices, which are called the **endpoints** of the edge, and are said to be **adjacent**. Two edges are **adjacent** if they have an endpoint in common. An edge which joins a vertex to itself is called a **loop**, while two edges which join the same pair of vertices are called **parallel**, or **multiple edges**. A pseudograph with no loops is called a **multigraph**, and a multigraph with no multiples edges is called a **graph**.*

[Warning: some authors use the term 'graph' in place of our 'pseudograph' or 'multigraph', and in place of our 'graph' use 'simple graph' or 'strict graph'. To make matters worse, some people use 'multigraph' to mean what we call a 'pseudograph'. Vertices are sometimes called 'points' or 'nodes', while edges are sometimes called 'lines' or 'arcs'.]

A graph is usually drawn with enlarged dots for the vertices, and straight lines (or sometimes curves) for edges, in such a way that a vertex and an edge are incident if and only if they meet in the diagram. We illustrate these concepts in Figs 2.1 and 2.2. We sometimes draw edges of a graph crossing each other, as in Fig. 2.3(a). This has no significance in graph theory.

Definition 2.2 *A **subgraph** consists of a subset of the vertices and a subset of the edges, with the property that for every edge in the subgraph, both its endpoints are in the subgraph. A **spanning** subgraph is one which contains all the vertices. The **induced** subgraph of G on a set W of vertices consists of W together with all the edges of G which join vertices in W. (This induced subgraph is also called the **subgraph generated by** W.)*

See Fig. 2.3 for an example of a graph G, together with a spanning subgraph and an induced subgraph. In a graph, as opposed to a multigraph or pseudograph,

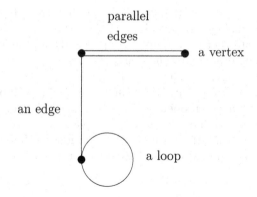

Fig. 2.1 The basic concepts.

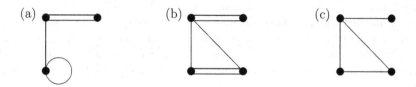

Fig. 2.2 Some examples. (a) A pseudograph, (b) a multigraph and (c) a graph.

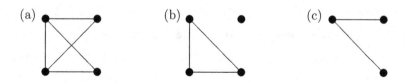

Fig. 2.3 A graph and some subgraphs. (a) A graph G, (b) a spanning subgraph and (c) an induced subgraph.

each edge is determined by its endpoints, and we shall often write uv for the edge joining the vertices u and v.

Definition 2.3 *A* **walk** *of* **length** n *is a sequence* $v_1 e_1 v_2 e_2 \cdots v_n e_n v_{n+1}$ *of vertices and edges such that each is incident to the next. It is* **closed** *if it ends up at the same place it started (i.e. if $v_1 = v_{n+1}$), and* **open** *otherwise.*

A **trail** *is a walk in which all edges are distinct. A* **circuit** *is a non-trivial closed trail, that is, a closed trail with at least one edge.*

A **path** *is a trail in which all the vertices are distinct (except possibly v_1 and v_{n+1}). A* **cycle** *is a circuit which does not contain a vertex twice (except at the beginning and end).*

Some authors omit the vertices in this definition, as they are determined by the edges (except in the trivial case of a walk in which all the edges go between the same two vertices). Note that on a graph, as opposed to a pseudograph or multigraph, the edges are determined by their endpoints, so a walk is completely specified by the sequence of vertices. Thus, we often write $v_1 v_2 \cdots v_n v_{n+1}$ as an abbreviation for the walk $v_1 e_1 v_2 e_2 \cdots v_n e_n v_{n+1}$ in a graph. For example, in the graph shown in Fig. 2.4, we have an open walk *abcbeg*, a closed walk *abcbegda*, an open trail *abcebad*, a closed trail or circuit *abcebafhgda*, and a cycle *abcegda*. Again, you should be aware that the terms used by various authors for these concepts vary widely, and the same word may have different meanings in different books.

Definition 2.4 *Two vertices are said to be* **connected** *if there is a walk from one to the other.*

Thus, in Fig. 2.5, the vertex v is connected to w, but not to x. It is intuitively obvious from such a picture that all the vertices connected to v are connected to each other. Similarly, if v is not connected to x, then v is not connected to any vertex which is connected to x. This is expressed mathematically by saying that connectedness is an equivalence relation on the vertices, which we now prove formally.

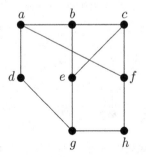

Fig. 2.4 A graph to illustrate Definition 2.3.

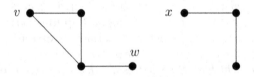

Fig. 2.5 Connectedness.

Lemma 2.5 *If u, v, and w are vertices in a graph (or pseudograph or multigraph), then*

1. *u is connected to u;*
2. *if u is connected to v then v is connected to u;*
3. *if u is connected to v and v is connected to w, then u is connected to w.*

Proof

1. The walk u connects u with itself.
2. If $ue_1v_1e_2v_2\cdots e_nv$ is a walk from u to v, then $ve_n\cdots v_2e_2v_1e_1u$ is a walk from v to u.
3. If $ue_1v_1e_2v_2\cdots e_nv$ is a walk from u to v, and $vf_1w_1f_2w_2\cdots f_mw$ is a walk from v to w, then $ue_1v_1e_2v_2\cdots e_nvf_1w_1f_2w_2\cdots f_mw$ is a walk from u to w. □

Thus, the vertices are partitioned into equivalence classes under this equivalence relation. There are no edges between a vertex v in one equivalence class and a vertex x in another, because otherwise v and x would be connected, which contradicts the assumption that they are in different equivalence classes. So we can obtain all the edges by looking at one equivalence class at a time.

Definition 2.6 *The induced subgraph on such an equivalence class is called a* **connected component** *or just* **component** *of the graph. A graph is* **connected** *if there is just one equivalence class, that is, if every pair of vertices is connected.*

For example, the graph in Fig. 2.5 has just two components, illustrated in Fig. 2.6.

Definition 2.7 *A* **tree** *is a connected graph with no cycles. A* **forest** *is a graph with no cycles.*

Examples are given in Fig. 2.7. The following characterization of trees will be needed at one point later on (in the proof of Theorem 8.19), but is not essential to the main theme of the book.

Lemma 2.8 *Let G be a connected graph with p vertices. Then, G is a tree if and only if G has $p - 1$ edges.*

Proof First, suppose that G is a tree with p vertices. We prove by induction on p that G has $p - 1$ edges. The induction starts with the trivial graph, with

Fig. 2.6 Components. (a) One component and (b) the other component.

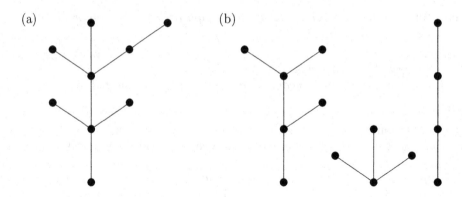

Fig. 2.7 Trees and forests. (a) A tree and (b) a forest.

one vertex and no edges. Now suppose $p > 1$, so that G has at least one edge (for otherwise G is not connected), and remove an edge from G. This disconnects G into two pieces, for otherwise G would have a circuit, which is a contradiction. Both pieces are trees, because they are connected and have no cycles. If they have k and l vertices, then by induction they have $k - 1$ and $l - 1$ edges, respectively, so the total number of edges in G is

$$(k - 1) + (l - 1) + 1 = (k + l) - 1$$
$$= p - 1$$

as required.

Conversely, suppose that G is not a tree. Then we can remove an edge from a circuit of G, and the resulting graph is still connected. Now continue removing edges in this way until there are no circuits left. At this point, the graph is a tree with p vertices, so has $p - 1$ edges. Therefore G has strictly more than $p - 1$ edges. $\square$

Definition 2.9 *The* **degree** *(or* **valence** *or* **valency***) of a vertex is the number of edges which are incident to it. (In a pseudograph, we usually count a loop twice.) The degree of a vertex v will be denoted $d(v)$.*

Notation We denote by $d(G)$ the average degree of the vertices of G. The minimum degree of the vertices of G is denoted $\delta(G)$, and the maximum degree by $\Delta(G)$.

Definition 2.10 *A graph is* **complete** *if every pair of vertices is adjacent. A graph (or multigraph) is* **bipartite** *if the vertices can be partitioned into two sets X and Y such that all the edges join a vertex in X to a vertex in Y. A graph is* **complete bipartite** *if it contains all possible edges from a vertex in X to a vertex in Y.*

The complete graph on n vertices is usually denoted K_n, while the complete bipartite graph on two sets of m and n vertices is denoted $K_{m,n}$. Figure 2.8(a) shows K_4 and Fig. 2.8(c) shows $K_{3,2}$.

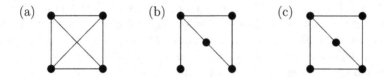

Fig. 2.8 Some more graphs. (a) A complete graph, (b) a bipartite graph and (c) a complete bipartite graph.

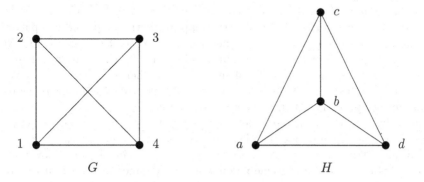

Fig. 2.9 A planar graph G and a plane graph H isomorphic to G.

Definition 2.11 *Two graphs (or pseudographs or multigraphs) G and H are* **isomorphic** *if there is a one-to-one correspondence between the vertices, and a one-to-one correspondence between the edges, which preserves incidence. In other words, if the edge e in G corresponds to the edge f in H, then the endpoints of e correspond to the endpoints of f. A graph (or pseudograph or multigraph) is* **plane** *if it is drawn in the plane with no two edges crossing each other, and is* **planar** *if it is isomorphic to a plane graph (or pseudograph or multigraph).*

In Fig. 2.9, G is a planar graph, on the vertices labelled $1, 2, 3, 4$, and H is a plane graph isomorphic to G, on the vertices labelled a, b, c, d. An isomorphism is given by the map taking 1 to a, 2 to b, 3 to c, and 4 to d. Another pair of isomorphic graphs is shown in Fig. 2.12.

2.2 Maps

In order to study the four-colour problem mathematically, it is first necessary to provide precise definitions of all the concepts involved. There are several ways of doing this, and as you might by now expect, there is little agreement about which is best. For our purposes, we shall consider a **map** M as consisting of a planar pseudograph $G(M)$, called the **underlying (pseudo)graph** of M, together with an embedding of M in the plane. By an **embedding**, we mean a drawing in which the edges do not cross.

In reality, this concept of map is still too general. When we think of drawing a map, with the lines (edges) corresponding to boundaries between countries, we only need to put vertices where three or more regions meet at a point. Vertices of degree 0 or 1 do not occur, and vertices of degree 2 can be eliminated. Thus, we may assume if we like that every vertex of $G(M)$ has degree at least 3. Unless otherwise specified, we shall make this assumption throughout. For technical reasons, when we consider the colouring of maps, we consider the exterior region of the plane to be one of the countries which needs to be coloured.

There are further simplifications we can make. For example, if $G(M)$ has a loop, then this loop divides the plane into an inside and an outside (note that there may be edges of the graph **inside** the loop!). On both the inside and the outside of the loop there is only one country which has this loop as a boundary edge. So if we can 4-colour the part of the map inside the loop, and the part of the map outside the loop, separately, then we can 4-colour the whole map. (All we have to do is ensure that the two countries bounded by the loop are coloured different colours—if they are both coloured red, say, then recolour red as green and green as red inside the loop.) Thus, we may assume that $G(M)$ has no loops, so is a multigraph rather than a general pseudograph.

Similarly, if $G(M)$ has two parallel edges, then these edges form a closed curve in the plane, and again divide the plane into an inside and an outside. This time we may have two (not necessarily distinct) colours inside, bounded by these edges, and two colours (or one colour) outside. Again, since we have four colours available, we can arrange that the four or fewer colours used by the countries adjacent to one or other of our two edges, are distinct. Thus, we may assume that $G(M)$ has no parallel edges, so is genuinely a graph rather than a multigraph.

We may also assume that $G(M)$ is connected. For if not, we remove one component C and colour the map which remains. We now only need to colour the map corresponding to C, and ensure that the colour of the outside country matches the colour of the country of M it came from. As usual, this may be achieved by changing the colours in one of the two maps we are putting together.

Note also that in any sensible map, no country has a boundary with itself. If it did, then removing an edge from this boundary would disconnect the graph. Conversely, if removing an edge from $G(M)$ would disconnect the graph, then the two countries in M on either side of this edge can be connected without crossing any boundaries. In other words, they were already one and the same country. Thus, we can exclude this case also. (An edge of a connected graph G whose removal disconnects G is called a **bridge**.)

It follows that for the purposes of considering the four-colour problem, and many similar problems, we may assume that $G(M)$ is a connected plane graph (no loops or parallel edges) with no bridges and no vertices of degree less than 3. A map M with this property is called a **standard map**. We shall often assume our maps are standard, without necessarily saying so explicitly each time.

2.3   Duality

You may have come across the idea of duality of regular polyhedra, in which you join up the midpoints of adjacent faces of one polyhedron to obtain the dual polyhedron. Doing this a second time gets you back to a smaller version of the original polyhedron. In this duality, a cube is dual to an octahedron, a tetrahedron is self-dual, and a dodecahedron is dual to an icosahedron.

A similar procedure can be performed on any polyhedron whatsoever, that is, any solid figure whose surface is made out of polygons. Indeed, with a little stretch of the imagination this procedure can be generalized to any connected map, or plane pseudograph. Thus given a map M, you draw a vertex of $D(M)$ in the interior of each region of M (including the exterior region), and join them by edges, one edge of $D(M)$ crossing each edge of M. The new map (or plane pseudograph) $D(M)$ is called the **dual** of M. See Fig. 2.10 for an example. Notice that, at least in this example, $D(D(M))$ is isomorphic to M.

Remark If we try to generalize to non-connected maps, then the dual is no longer well-defined. Moreover, it is obvious that the dual of any (connected or non-connected) map is always connected, so in this case $D(D(M))$ cannot be isomorphic to M.

We now ask, how do the properties of $D(M)$ relate to the properties of M? We have already seen that if M has p vertices, q edges and r faces, then $D(M)$ has r vertices and q edges. To show that $D(M)$ really is a 'dual' to M, it is necessary to show that $D(D(M))$ is isomorphic to M, and in particular that $D(M)$ has p faces.

To see this, note that a vertex v of degree d in M is surrounded by d regions, each adjacent to two others. Thus, in $D(M)$ we obtain a cycle of vertices, bounding a region which contains v and (because M is connected) no other vertex of M. Moreover, every region of $D(M)$ has edges of M extending into it, so contains at least one vertex of M. So in constructing $D(D(M))$, we might as well take the vertices to be the vertices of M, and similarly for the edges (see Fig. 2.11).

This means that a vertex of degree d in M corresponds to a region with d sides in $D(M)$, and vice versa. In particular, a vertex of degree 1 corresponds to

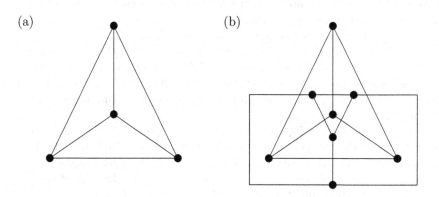

Fig. 2.10 Duality. (a) A map M and (b) M with its dual.

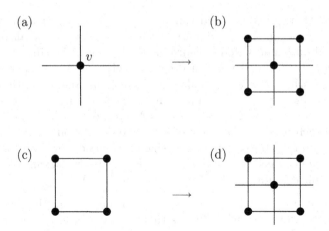

Fig. 2.11 Correspondence between vertices and faces. (a) A vertex in M, (b) the corresponding region in $D(M)$, (c) a region in $D(M)$ and (d) the corresponding vertex in $D(D(M))$.

a loop with nothing in its interior. More generally, any loop in M corresponds to a bridge in the dual $D(M)$. Similarly, a pair of parallel edges in M corresponds to a pair of edges in $D(M)$ whose removal disconnects $D(M)$.

We can think of the dual graph of a map as consisting of vertices representing the capital cities of all the countries, and edges representing roads between these cities, one road crossing each segment of border between two adjacent countries. The dual graph gives us an easier way to give the four-colour problem a precise mathematical form. Rather than colouring the countries, we can consider colouring the capital cities—this is of course equivalent, since the capital cities are in one-to-one correspondence with the countries themselves. The condition that no two adjacent countries should have the same colour now translates into saying that two capital cities connected by a direct road should not have the same colour. Equivalently, we are trying to colour the vertices of $D(M)$ in such a way that adjacent vertices have different colours.

Now it is clear that we need to ignore loops, as it is impossible to colour a vertex differently from itself. Also, if two vertices are adjacent, then it does not matter how many edges there are joining them. Thus we can simplify multiple edges to single edges. In other words, we might as well restrict to graphs. It is also clear that if we can prove the four-colour conjecture for connected graphs, then we can prove it for all graphs.

The concept of duality can be applied more generally, to any plane connected psuedograph. Notice that a given planar graph can in general be embedded in the plane in many different ways, and of course the four-colour conjecture does not depend on the particular planar embedding used. However, the **interpretation** of the graph as the dual graph of a map may be completely different.

To put this another way, different plane embeddings of a planar graph G may give different dual graphs $D(G)$. In other words, we can have isomorphic graphs

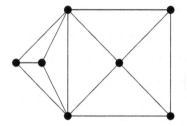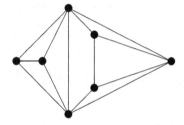

Fig. 2.12 Two isomorphic graphs with non-isomorphic duals.

with non-isomorphic duals (see Fig. 2.12). Here, the outer face of the first graph has five edges, so the dual graph has a vertex of degree 5. On the other hand, all faces of the second graph have three or four edges, so all vertices of the dual graph have degree 3 or 4. Therefore the two dual graphs cannot be isomorphic.

2.4 Euler circuits

Euler (pronounced 'oiler') was a famous eighteenth century Swiss mathematician. He is generally credited with 'inventing' graph theory in his paper on the Königsberg Bridge problem in 1736, although he did not draw any graphs, or use the word 'graph'. Indeed, the use of the word graph in this context only goes back to Sylvester in 1878.

Simply described, the problem is whether it was possible to walk around the city of Königsberg, crossing each of the seven bridges exactly once, and arrive back at one's starting point. Figure 2.13 gives a schematic representation of the bridges over the River Pregel at that date. A process of trial and error will soon convince you that no such walk is possible. Euler's contribution was to produce a rigorous proof of this, and to generalize it to a criterion for deciding, given any arrangement for bridges, whether such a walk exists. He proved the necessity of his condition, but did not apparently see the need to prove sufficiency.

The idea is to label the four land areas (the two islands and the two banks of the river) with the letters, A, B, C, D. Then, except at the beginning and the end of the walk, every time you enter and leave one of the areas A, B, C, D, you do so by crossing one bridge to enter, and another bridge to leave. So you use up two bridges at a time, and therefore you use an even number of bridges altogether. Unfortunately, each of the four areas A, B, C, D has an odd number of bridges to it, so the walk is impossible. Indeed, the starting point is no different from any of the others: you use one bridge to leave initially, then two for each time you return and leave again, and finally one more bridge to return for the last time. So, again there should be an even number of bridges from that point.

We can express this in the language of graph theory by replacing the land areas by vertices, and the bridges between them by edges. In this way we obtain a multigraph, as in Fig. 2.14. Corresponding to the fact that each area A, B, C, D has an odd number of bridges to it, is the fact that each of the vertices of the

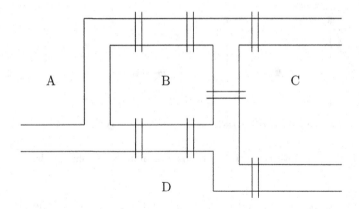

Fig. 2.13 The Königsberg bridges.

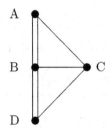

Fig. 2.14 The multigraph of the Königsberg bridges.

multigraph has odd degree. The walk that we are looking for is a circuit which uses each edge exactly once. Such a circuit in any multigraph is now called an **Euler circuit**. The above discussion essentially contains Euler's proof of the following theorem, although he did not use the modern language of graph theory.

Theorem 2.12 *A multigraph possesses an Euler circuit only if the degree of every vertex is even, and every edge is in the same component (in other words, it is connected except for isolated vertices).*

Notice that we have not proved the converse of this result! It is not as easy as you may think. The problem is that if you just keep walking you might run out of places to go to from where you are, while still not having used all the edges in some other part of the multigraph. Thus, you may have to go back and insert an extra circuit into the walk somewhere. Formally, we need to express this as a proof by induction, or the following contrapositive version (a proof by contradiction).

Theorem 2.13 *A multigraph possesses an Euler curcuit if it is connected and the degree of every vertex is even.*

Proof Suppose G is a connected multigraph and the degree of every vertex is even. If G does not possess an Euler circuit, then let $v_1e_1v_2e_2\cdots v_ne_nv_1$ be a circuit which is as long as possible subject to not containing any edge more than once. Call this circuit C. By assumption, there is an edge e not in C. Since G is connected, we may assume that one of the endpoints of e is already in C, so let this endpoint be v_i, and let the other endpoint of e be w_1. Now an even number (possibly 0) of the edges incident with w_1 are already used in C. Also e is incident with w_1, and w_1 has even degree, so there is another edge, f_1, say, incident with w_1 and not used in C.

Let w_2 be the other endpoint of f_1. The same argument now applies to w_2, and by induction we obtain a walk

$$v_iew_1f_1w_2f_2\cdots$$

such that all the edges $e, f_1, f_2, \ldots$ are not used in C. Since the multigraph is finite, this process must eventually stop, but the only way this can happen is if we arrive back at a vertex after using all edges incident with it, that is, an **even** number of edges. But the only vertex where this can happen is v_i, so we obtain a circuit

$$v_iew_1f_1w_2f_2\cdots w_kf_kv_i$$

and then

$$v_1e_1v_2e_2\cdots e_{i-1}v_iew_1f_1w_2\cdots w_kf_kv_ie_i\cdots v_ne_nv_1$$

is a longer circuit than C, with no repeated edges. This contradiction completes the proof. □

Exercises

Exercise 2.1 Let G be the graph in Fig. 2.15. Draw the induced subgraphs on the following sets of vertices.

1. $\{b, c, e, f\}$;
2. $\{b, c, f, g\}$;
3. $\{a, d, e, f\}$.

Draw a spanning subgraph which is a tree.

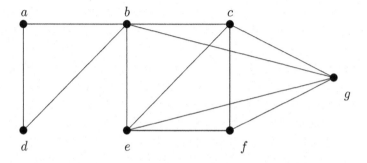

Fig. 2.15 A graph for Exercise 2.1.

Exercise 2.2 We say that two plane graphs are **isomorphic as plane graphs** (as opposed to being isomorphic only as graphs) if there are one-to-one correspondences between vertices, edges and faces, preserving incidence. A plane graph is called **self-dual** if it is isomorphic to its dual graph, as plane graphs. Show that the tetrahedron (i.e. the complete graph K_4 on four vertices) is self-dual. Find another plane graph which is self-dual.

Exercise 2.3 Prove formally that the two graphs in Fig. 2.12 are isomorphic (as graphs), but are not isomorphic as plane graphs.

Exercise 2.4 Prove that a standard map can be 2-coloured if and only if every vertex of the underlying graph has even valency.

Exercise 2.5 Prove that the following three conditions on a graph G are equivalent:
1. G is bipartite;
2. all cycles in G have even length;
3. the vertices of G can be 2-coloured, so that adjacent vertices have different colours.

Exercise 2.6 Show that if M has a cutvertex (i.e. a vertex whose removal disconnects M) then so does $D(M)$.

Exercise 2.7 Show that every edge in a tree is a bridge.

Exercise 2.8 Prove that if G is a connected graph such that every edge is a bridge, then G is a tree.

Exercise 2.9 Show that there are just four non-isomorphic graphs on three vertices.

Exercise 2.10 Show that there are just 11 non-isomorphic graphs on four vertices. How many of these are connected? How many of these are trees?

Exercise 2.11 Show that there are just six non-isomorphic trees on six vertices.

3
Applications of Euler's formula

3.1 Euler's formula

You may have met Euler's formula for regular polyhedra: if V is the number of vertices, E the number of edges, and F the number of faces, then $V - E + F = 2$. There are five such regular polyhedra, and you can check this equation in the five cases (see Table 3.1). Moreover, since this is a purely combinatorial result, which does not depend on the actual shapes of the polyhedra, we can stretch them and flatten them out onto a piece of paper, and hence draw them as maps or graphs in the plane. These graphs are drawn in Figs 3.1 and 3.2. Notice that one of the faces of the polyhedron has now become the whole of the outside region of the map.

What is not so well known is that Euler's formula actually holds for any connected map, or, equivalently, for any connected plane graph, or even connected plane pseudograph. Euler announced this result in 1750, but admitted he could not prove it. His proof published in 1752 does not actually cover all cases, and the first complete proof was given by Cauchy in 1813.

We prove the theorem by induction on the number of edges in the graph. First, we need a technical lemma, which is intuitively 'obvious' but is nevertheless worth proving carefully.

Lemma 3.1 *If a pseudograph has at least one edge, and has no vertex of degree 1, then it contains a cycle.*

Proof Let $e_1 = v_1 v_2$ be an edge. Then v_2 does not have degree 1, so either e_1 is a loop, in which case we are done, or there is a second edge $e_2 = v_2 v_3$, say,

Table 3.1 Euler's formula for the five Platonic solids

	V	E	F	$V - E + F$
Tetrahedron	4	6	4	2
Cube	8	12	6	2
Octahedron	6	12	8	2
Dodecahedron	20	30	12	2
Icosahedron	12	30	20	2

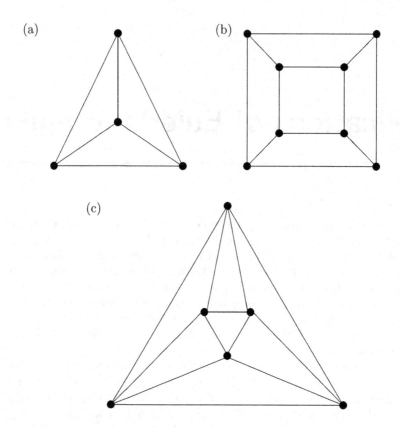

Fig. 3.1 The underlying graphs of the five Platonic solids I. (a) Tetrahedron, (b) cube and (c) octahedron.

incident with v_2. Thus there is a trail $v_1 e_1 v_2 e_2 v_3$. Again, v_3 does not have degree 1, so (unless $v_3 = v_2$ or v_1, in which case we are done) there is another edge $e_3 = v_3 v_4 \neq e_2$ extending the trail to $v_1 e_1 v_2 e_2 v_3 e_3 v_4$. In general, we obtain a trail $v_1 e_1 v_2 e_2 \cdots e_{i-1} v_i$, and v_i does not have degree 1, so we can adjoin another edge $e_i = v_i v_{i+1}$ to the trail. Eventually, since the pseudograph is finite, we must end up at a vertex we have already had. Suppose that the first time this happens is when we try to adjoin v_{i+1}, and we find that $v_{i+1} = v_j$ for some $j \leqslant i$. Then $v_j e_j v_{j+1} \cdots v_i e_i v_{i+1}$ is a cycle, since all the vertices $v_j, v_{j+1}, \ldots, v_i$ are distinct. □

Theorem 3.2 (Euler's theorem) *If G is a connected plane pseudograph, with p vertices, q edges, and r faces, then $p - q + r = 2$.*

Proof By induction on the number q of edges. If there is a vertex of degree 1, remove it and the edge incident to it (see Fig. 3.4(a)). This does not change the number of faces, so does not change the value of $p - q + r$. If there is no vertex of degree 1, then as in Lemma 3.1 we can keep walking along the pseudograph until

(a)

(b)

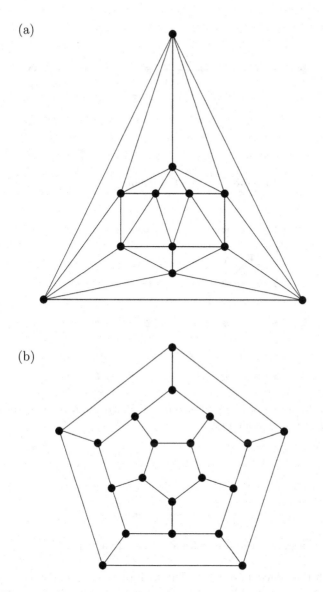

Fig. 3.2 The underlying graphs of the five Platonic solids II. (a) Icosahedron and (b) Dodecahedron.

we get to somewhere we have been before. Therefore, there is a circuit in the pseudograph, and we can remove one of its edges while keeping the pseudograph connected (see Fig. 3.4(b)). This does not change the number of vertices, but decreases the number of edges and faces by 1 (since two faces have been combined into one), and therefore, the value of $p - q + r$ is again unchanged. Eventually, we

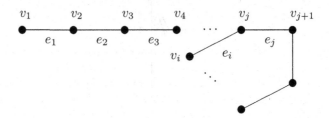

Fig. 3.3 The proof of Lemma 3.1.

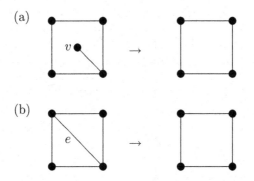

Fig. 3.4 Steps in the proof of Euler's theorem. (a) Removing a vertex v of degree 1 and (b) removing an edge e from a cycle.

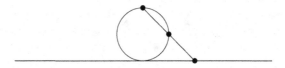

Fig. 3.5 Projecting from a sphere to a plane.

have removed all the edges, and we are left with one vertex (since the resulting pseudograph is still connnected) and one face, so $p - q + r = 2$. □

As we saw when looking at the regular polyhedra, this formula really holds on the surface of a sphere: just make a hole in the middle of some region, and flatten out the sphere onto a plane. This is, of course, what we do when making real maps of the surface of the Earth. In particular, all questions and results about plane graphs can be reformulated in terms of graphs drawn on the surface of a sphere. More formally, we can project a map from a sphere onto a plane (or vice versa) by drawing rays from the North Pole onto a horizontal plane through the South Pole (see Fig. 3.5).

3.2   Applications

From now on we shall usually use p for the number of vertices of a pseudograph G, and q for the number of edges. For plane pseudographs we also use r for the number of faces (including the exterior face). There are some useful inequalities involving p, q, and r for plane graphs, most of which follow from Euler's formula and a simple counting argument. In some cases, but by no means all, these generalize to multigraphs or pseudographs.

The basic counting argument is the so-called 'handshaking lemma', which says that if a number of people shake hands, then the total number of hands being shaken (counted with multiplicities) is even. This is obvious because each individual handshaking involves two hands. It follows that the number of people who have shaken hands an odd number of times is even. More generally, if you take the sum over all people, of the number of hands they have shaken, you get twice the total number of handshakings. This result first appears in Euler's paper of 1736, mentioned in Section 2.4. In graph-theoretical language, we have the following.

Lemma 3.3 *The sum of the degrees of the vertices of a pseudograph is equal to twice the number of edges. Writing p_i for the number of vertices of degree i, this can be expressed*

$$\sum_{i=1}^{\infty} ip_i = 2q.$$

Proof First note that this sum is really a finite sum, as there are no vertices of degree more than q. Now divide each edge into two half-edges. Thus the total number of half-edges is $2q$. On the other hand, each half-edge is incident with a unique vertex, and the number of half-edges incident with a vertex is exactly the degree of that vertex. Therefore, the total number of half-edges is the sum of the degrees of the vertices.   □

There is a dual version of this for plane pseudographs, obtained by counting the edges around the faces. Since each edge is again counted twice, we have the following.

Lemma 3.4 *If G is a plane pseudograph, and r_i is the number of faces with i sides, then*

$$\sum_{i=1}^{\infty} ir_i = 2q.$$

In this formula, an edge is counted twice if it occurs twice in the boundary walk of a face. Thus, for example, in Fig. 3.6 face A has 7 sides, face B has 12 sides, face C has 8 sides, and face D has one side.

The following are simple corollaries of the two forms of the handshaking lemma.

Corollary 3.5 *For any pseudograph G, the average degree of the vertices is $2q/p$.*

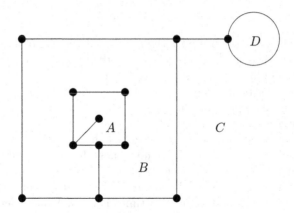

Fig. 3.6 Counting edges of unusual faces. Faces A, B, C, and D have 7, 12, 8, and 1 edge(s), respectively.

Corollary 3.6 *For any plane pseudograph G, the average number of sides of the faces is $2q/r$.*

Many useful results about planar graphs can be obtained by combining the handshaking lemmas with Euler's formula. For example,

Proposition 3.7 *In any plane graph in which all faces are triangles, $q = 3p - 6$.*

Proof Since every face (including the exterior face) has three edges, and every edge belongs to two faces, the handshaking lemma says $2q = 3r$. Substituting into Euler's equation gives $6 = 3p - 3q + 3r = 3p - q$, that is $q = 3p - 6$. $\quad\square$

Theorem 3.8 *In any planar graph with at least 3 vertices, $q \leqslant 3p - 6$.*

Proof First, embed the graph in the plane, and then add edges until we have a connected graph G, which contains a cycle. This is possible, since there are at least 3 vertices. If $q \leqslant 3p - 6$ for the new graph G, then it is certainly true for the original graph. Now G has at least two faces, and so at least 3 sides to each face. Therefore $3r \leqslant 2q$, so by Euler's formula

$$6 = 3r - 3q + 3p$$
$$\leqslant 3p - q$$

whence $q \leqslant 3p - 6$ as required. $\quad\square$

The above results can be used in certain circumstances to prove a graph is not planar. For example, the **complete graph** on n vertices, written K_n, is defined by joining all pairs of vertices by an edge, and we show that K_5 is not planar (see Fig. 3.7). For K_5 has 5 vertices and 10 edges, so $p = 5$ and $q = 10$, and so $q \nleqslant 3p - 6$. As a corollary we have that K_n is non-planar for every $n \geqslant 5$. On the other hand, K_4 is planar—it is isomorphic to the tetrahedron (see Fig. 3.1(a)).

Remark It is sometimes said that Möbius originated the four-colour problem in 1840. In fact, according to May [36], Möbius' remark was equivalent to the

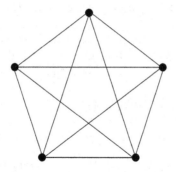

Fig. 3.7 The complete graph K_5.

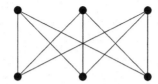

Fig. 3.8 The 'utilities graph' $K_{3,3}$.

statement that K_5 is not planar. It is a common misconception (shared apparently even by De Morgan in 1852) that this implies the four-colour theorem. However, the fact that a planar graph does not contain a subgraph K_5 merely means that there is no **local** obstruction to a 4-colouring of the vertices. There may still be a **global** obstruction to 4-colouring, and this is the crux of the problem.

We have defined the **complete bipartite graph** $K_{m,n}$ by taking one set of m vertices and another set of n vertices, and joining every vertex in the first set to every vertex in the second set. Thus $K_{m,n}$ has $m+n$ vertices and mn edges. Then $K_{3,3}$ is non-planar, as we shall see. This graph is sometimes called the 'utilities graph': three houses are to be connected to three utilities: gas, electricity and water. Can this be done in such a way that no pipes or cables cross? The graph is as in Fig. 3.8 and a little thought will convince you that it is not possible to draw this without crossing edges.

In fact, this can also be proved from Euler's formula: the shortest cycles in the graph have length 4 (all cycles in a bipartite graph have even length, since they must alternate between the two sets of vertices), so if it were planar, all the faces would have to have at least four edges. This implies that $2q \geqslant 4r$, that is, $q \geqslant 2r$, and substituting into Euler's formula gives $4 = 2p - 2q + 2r \leqslant 2p - 2q + q$, that is $q \leqslant 2p - 4$. However, in the graph $K_{3,3}$, we have $2p - 4 = 8$, but $q = 9$. Thus $q \nleqslant 2p - 4$, so the graph cannot be planar. It follows that $K_{m,n}$ is non-planar whenever $m \geqslant 3$ and $n \geqslant 3$. On the other hand, $K_{2,n}$ is planar for any n (prove it!).

We define the **girth** of a graph to be the length of the shortest cycle in the graph. Note that this is only defined for graphs which do contain cycles. Thus, K_5 has girth 3, while $K_{3,3}$ has girth 4. With this definition, we can generalize Theorem 3.8.

Theorem 3.9 *In any plane graph of girth l,*

$$q \leqslant \frac{l}{l-2}(p-2).$$

Proof First add enough edges to make the graph connected, without changing the girth. Then every face has at least l sides, so the handshaking lemma says $2q \geqslant lr$. We now substitute $r \leqslant 2q/l$ into Euler's formula $p - q + r = 2$, giving

$$
\begin{aligned}
p - 2 = q - r \\
\geqslant q - \frac{2q}{l} \\
\geqslant \frac{l-2}{l}q
\end{aligned}
$$

and hence

$$q \leqslant \frac{l}{(l-2)}(p-2)$$

as required. $\qquad\square$

Definition 3.10 *The **Petersen graph** may be defined by taking 10 vertices corresponding to the unordered pairs from the numbers $1, 2, 3, 4, 5$, and joining two vertices when the corresponding pairs have no number in common.*

Corollary 3.11 *The Petersen graph is not planar.*

Proof By inspection, the graph (see Figs 3.9 and 3.10) has girth 5, and has 10 vertices and 15 edges, but $15 \not\leqslant \frac{5}{3} \times 8$, so the graph is not planar. $\qquad\square$

We have proved various corollaries of Euler's formula, and deduced that certain graphs are non-planar. It is clear that if G is a non-planar graph, and G is a subgraph of a graph H, then H is non-planar. But we can actually do better than this. For example, the Petersen graph contains a subgraph which is a 'subdivision' of $K_{3,3}$, and this shows that the Petersen graph is non-planar. We will define the term **subdivision** precisely later (see Section 7.1), but for the moment we just think of it as repeatedly adding vertices in the middle of existing edges. It is, then, clear that subdividing a graph does not change the property of being planar. This proves

Theorem 3.12 *A graph G is planar only if it contains no subgraph which is a subdivision of K_5 or $K_{3,3}$.*

The converse of this theorem, known as Kuratowski's theorem, is much harder, and we will prove it in Section 7.2. For now, we prove some more corollaries of Euler's theorem which will be useful in our analysis of the four-colour problem. The most important one from our point of view is the following.

Corollary 3.13 *Every plane connected pseudograph G with all vertices of degree at least 3 has a face with at most five sides.*

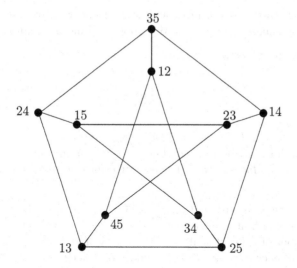

Fig. 3.9 The Petersen graph.

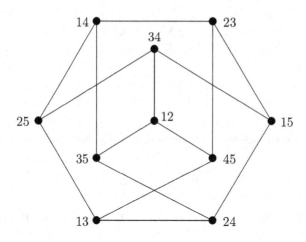

Fig. 3.10 Another drawing of the Petersen graph.

Proof By contradiction. Suppose that all the vertices have degree at least 3, and all the faces have at least six sides. Then the usual handshaking lemmas imply that $2q \geqslant 3p$ and $2q \geqslant 6r$, so substituting in Euler's formula gives

$$2 = p - q + r$$
$$\leqslant \frac{2}{3}q - q + \frac{1}{3}q$$
$$= 0$$

which is the desired contradiction. □

We can strengthen this result in various ways. First, we have a lemma which uses only the handshaking lemma, and does not use Euler's formula.

Lemma 3.14 *In any plane pseudograph,*

$$\sum_{i=1}^{\infty}(6-i)r_i = 6r - 2q,$$

where r_i denotes the number of i-sided faces (regions).

(In a multigraph there are no 1-sided faces, so the summation starts at $i = 2$. In a graph there are no faces with just one or two sides, so the summation starts with $i = 3$.)

Proof First note that the summation $\sum_{i=1}^{\infty}(6-i)r_i$ is really a finite sum, since the number of faces is finite. We observe that $\sum_{i=1}^{\infty} ir_i$ is the sum over all the faces, of the number of sides of that face. In other words, it counts all the edges exactly twice, since each edge is on the boundary of exactly two faces. (In certain cases, these two faces may actually be the same face.) Thus $\sum_{i=1}^{\infty} ir_i = 2q$, as in Lemma 3.4 , and similarly $\sum_{i=1}^{\infty} r_i = r$, the total number of faces. Therefore,

$$6r - 2q = 6\sum_{i=1}^{\infty} r_i - \sum_{i=1}^{\infty} ir_i$$
$$= \sum_{i=1}^{\infty}(6-i)r_i.$$

$\square$

Proposition 3.15 *In any plane connected pseudograph G with all vertices of degree at least 3,*

$$\sum_{i=1}^{\infty}(6-i)r_i \geqslant 12.$$

(Again we can start the summation at $i = 2$ if G is a multigraph, and at $i = 3$ if G is a graph.)

Proof Since the vertices have degree at least 3, we have $2q \geqslant 3p$, so $6p - 4q \leqslant 0$. From Euler's formula, we have

$$12 = 6p - 6q + 6r$$
$$= (6p - 4q) + (6r - 2q)$$
$$\leqslant 6r - 2q$$

so $6r - 2q \geqslant 12$. Now apply Lemma 3.14.

$\square$

Corollary 3.16 *Every plane connected pseudograph G with all vertices of degree at least 3, and no faces with fewer than five sides, has at least 12 faces with five sides.*

Proof The terms of the sum $\sum_{i=1}^{\infty}(6-i)r_i$ are all negative for $i > 6$, and zero for $i = 6$, so for the sum to be positive, at least one of the r_i for $i \leqslant 5$ must be non-zero. If moreover $r_i = 0$ for $i < 5$, we obtain $r_5 \geqslant 12$. $\qquad\square$

The above results are stated in the form required for the face-colouring version of the four-colour conjecture. They all have 'dual' forms using vertices instead of faces, but you must be careful to get the conditions on the graphs correct. In particular, duality only really makes sense for **connected** plane graphs, so you need to consider separately the question as to whether the result remains true for non-connected graphs. We state them here, and leave the proofs as exercises. In each case we have given the statement for **plane** graphs, but since the actual planar embedding is irrelevant, the results are true for **planar** graphs.

Lemma 3.17 *In any plane pseudograph,*

$$\sum_{i=1}^{\infty}(6-i)p_i = 6p - 2q,$$

where p_i denotes the number of vertices of degree i.

Proposition 3.18 *In any plane, connected graph with at least three vertices,*

$$\sum_{i=1}^{\infty}(6-i)p_i \geqslant 12,$$

where p_i denotes the number of vertices of degree i. Equivalently,

$$\sum_{v}(6 - d(v)) \geqslant 12,$$

where the sum is taken over all vertices v. The same inequalities are true without the connectedness condition.

Corollary 3.19 *Every plane connected graph has a vertex with degree at most 5. Moreover, if there is no vertex with degree less than than 5, then there are at least 12 vertices of degree 5.*

The same is true without the connectedness condition.

Exercises

Exercise 3.1 For each of the five regular polyhedra, determine the minimum number of colours required to colour the faces.

Exercise 3.2 Prove that if G is a plane graph with p vertices, q edges, and r faces, and with exactly k connected components, then $p - q + r = k + 1$.

Exercise 3.3 Find a subgraph of the Petersen graph which is a subdivision of $K_{3,3}$.

Exercise 3.4 Suppose that G is a connected plane pseudograph, and for each i, let p_i be the number of vertices of degree i. Similarly, let r_i be the number of i-sided faces. Use Euler's formula to show that

$$\sum_{i=1}^{\infty}(4-i)(p_i+r_i) = 8.$$

Deduce that G has either a vertex of degree at most 3, or a face with at most three sides (or both).

Exercise 3.5 Show that in any standard map (i.e. connected plane graph with no bridges, and with all vertices of degree at least 3), the average number of neighbours of all regions is less than 6.

Exercise 3.6 Show that in any planar graph the average degree of the vertices is less than 6.

Exercise 3.7 Draw a picture which makes it obvious that $K_{2,n}$ is planar for every n.

Exercise 3.8 Prove Lemma 3.17, Proposition 3.18 and Corollary 3.19.

Exercise 3.9 Show that Theorem 3.8 is false for multigraphs. Where does the proof break down?

Exercise 3.10 (Möbius's problem) A certain king had five sons, and on his deathbed wanted to divide his kingdom between all his sons, in such a way that each son's kingdom bordered each of the others. What advice would you give to the king?

4

Kempe's approach

4.1 The first 'proof' of the four-colour theorem

Kempe (pronounced 'kemp') published his 'proof' of the four-colour theorem in 1879, and for a decade it was accepted as a valid proof, and an ingenious solution to the problem. It contained several clever ideas, which we now present. For historical reasons, we describe his arguments in the face-colouring form, as Kempe himself did. He also mentioned duality and the translation of the problem into the vertex-colouring form. We discuss this translation in Section 4.4.

We first define a **Kempe chain** to be the largest set of countries you can get to from a given place by keeping to countries of a particular two colours, and crossing at edges, not vertices. In other words, in the dual form it should consist of a largest connected subgraph consisting of the vertices coloured in the two chosen colours. For example, a red–green chain would be a largest 'connected' piece of the map which consists entirely of countries coloured red or green. Note that, in general, it need not look like a chain at all.

Lemma 4.1 *Let M be a 4-colourable map. If four countries meet at a point v, then the map can be 4-coloured in such a way that only three colours are used for these four countries.*

Proof Suppose you used four colours, say red, green, blue, yellow in order round the point. Then if the red and blue countries do not belong to the same red–blue chain, we can swap red and blue in **one** of these two chains, and obtain the required colouring. If they **do** belong to the same red–blue chain, then the green and yellow countries are separated from each other by this red–blue chain, and so they cannot belong to the same green–yellow chain. Therefore, we can swap the colours in one of these two green–yellow chains. □

The above proof is illustrated in Fig. 4.1, where a portion of a typical map is shown in each of the two cases. The colours are denoted by the letters R, G, B, Y, and the letters in brackets denote the colours after they have been changed.

Lemma 4.2 *Let M be a 4-colourable map. If five countries meet at a point v, then the map can be 4-coloured in such a way that only three colours are used for these five countries.*

31

(a)

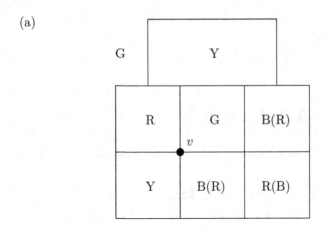

(b)

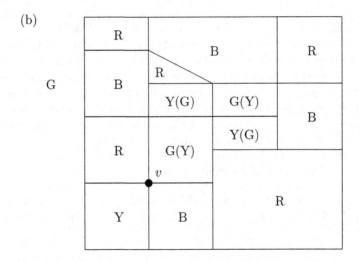

Fig. 4.1 The proof of Lemma 4.1. (a) Case 1: no R–B chain and (b) case 2: an R–B chain.

Proof Actually, this is the bit that Kempe got wrong. See if you can spot the error:

If you need all four colours around v, then the two countries coloured the same are separated by one country in one direction, and two in the other, so we may assume that the colours are red, green, blue, yellow, green in clockwise order. Now if the red and blue countries do not belong to the same red–blue chain, then we can interchange the colours in one of these chains, and so use only three colours. Similarly, if the red and yellow countries do not belong to the same red–yellow chain, we can again reduce to three colours.

The only other case (see Fig. 4.2) is where there is a red–blue chain isolating one green country, and a red–yellow chain isolating the other green country.

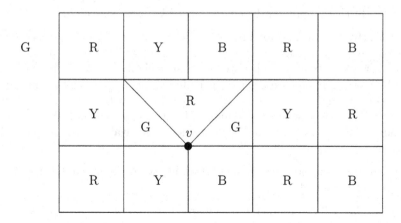

Fig. **4.2** The 'proof' of Lemma 4.2.

Thus, we can interchange the colours in the green–yellow chain containing the first of these green countries, and in the green–blue chain containing the second. This makes the colours red, yellow, blue, yellow, blue, clockwise in order, thus achieving the desired 3-colouring around this point. □

Did you spot the error? Don't worry if you did not—it took the world over ten years to spot it originally! We shall explain the problem in the next section. Meanwhile, keep thinking.

Kempe's 'proof' of the theorem now goes by induction on the number of countries. By Corollary 3.13, we can choose a country with five or fewer edges, and carve it up equally amongst its neighbours, to obtain a map with fewer countries, which can be four-coloured, by induction. Then we put the country back, and choose a colour for it. If it has three or fewer neighbours, then these use up at most three colours, so there is always one left. Similarly, if there are four neighbours, then Lemma 4.1 says you only need three colours for them, so there is still one left. And Lemma 4.2 would deal with the only remaining case, of five neighbours, if only the Lemma were true!

Remark Whilst we have noted that Kempe's argument was fallacious, we must still give him credit for several clever ideas. In particular, the general inductive argument, the use of Euler's formula to focus attention on a local area, and the Kempe-chain argument, are all essential ingredients of the eventual proof nearly 100 years later.

4.2 The five-colour theorem

Before we consider why Kempe's attempt at a proof fails, let us show how Heawood salvaged the five-colour theorem from the wreckage. The inductive argument and Euler's formula alone, without any Kempe chain argument, only allow one to prove the six-colour theorem, as follows.

Theorem 4.3 *Every map can be coloured with at most six colours.*

Proof Clearly, this is true for any map with no more than six countries, so suppose our map has at least seven countries. Then by Corollary 3.13 to Euler's formula, there is a country with fewer than six neighbours. If we remove this country by sharing it out amongst its neighbours, then the resulting map has one fewer country and so (by induction) can be coloured with at most six colours. Now replace the country which was removed. It has at most five neighbours which are already coloured, which leaves at least one colour available to complete the map colouring. □

Now to prove Heawood's five-colour theorem, we use Lemma 4.4 instead of the discredited Lemma 4.2.

Lemma 4.4 *Let M be a 5-colourable map. If five countries meet at a point, then the map can be 5-coloured in such a way that only four colours are used for these five countries.*

Proof For the sake of argument suppose that the five colours that you need are called red, orange, yellow, green and blue in cyclic order. If the red and yellow countries are not in the same red–yellow chain, then as before we swap the colours in one of these chains, and achieve the desired colouring. But if they are in the same red–yellow chain, then this separates the orange and green countries from each other, which are, therefore, not in the same orange–green chain, so one of them can be re-coloured. □

Theorem 4.5 (Heawood, 1890) *Every plane map can be coloured with at most five colours.*

Proof By Corollary 3.13 we can choose a country with five or fewer edges, and carve it up equally amongst its neighbours, to obtain a map with fewer countries, which can be 5-coloured, by induction. Then we put the country back, and choose a colour for it. If it has four or fewer neighbours, then these use up at most four colours, so there is always one left. Similarly, if there are five neighbours, then Lemma 4.4 says you only need four colours for them, so there is still one left. □

Where does Kempe's proof fail? It fails because you may not be able to change colours in both Kempe chains simultaneously. That is, by changing colours inside one chain, you completely change the other chain, so that changing colours inside this second chain no longer achieves the result it was supposed to achieve. This can happen if the two chains intersect each other—which they can do, since they have one colour in common.

Figure 4.3 shows a small example, with the colours labelled R, G, B, Y. Kempe's argument first swaps the colours in the G–Y chain containing the country 1. This has the effect of breaking the R–Y chain, and creating a G–B chain from country 2 round to the country 3. If we then perform the second colour-change required by Kempe, we swap the colours in this G–B chain, which has the effect of colouring country 3 with colour G. Thus, we have not achieved the

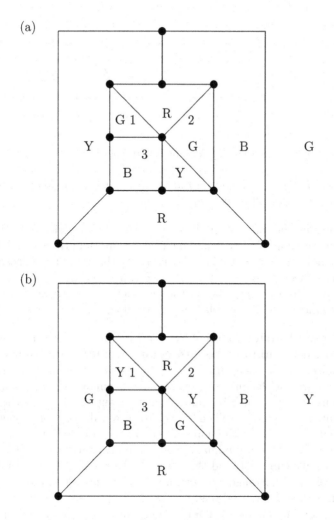

Fig. 4.3 A counterexample to Lemma 4.2. (a) The original colouring and (b) after the G–Y colour interchange.

desired result, of colouring the five countries around the central vertex with three colours.

4.3 A reduction theorem

As we have seen, Kempe's argument (amended by Heawood) does actually prove the five-colour theorem. If we want to prove the four-colour theorem, we have to be much more careful. First, we give a useful **reduction theorem**, which was first proved by Cayley a year before Kempe's paper. A graph or map is called **cubic** (or **trivalent**) if every vertex has degree 3.

Fig. 4.4 Making a map cubic.

Theorem 4.6 *If the (face-colouring) four-colour theorem holds for cubic plane maps, then it holds for all plane maps.*

Proof Certainly, the four-colour theorem implies this restricted version. Conversely, given any map, if there is a point where more than three countries meet, introduce a small country centred on this point, so that round its edges now only three countries meet at any one point (see Fig. 4.4). Then if the new map can be coloured with four colours, so can the old one: we simply delete these small introduced countries, with no violation of the colouring conditions. □

Now the general strategy of proof for a theorem like the four-colour theorem is often to consider a **minimal counterexample**. That is, we assume that the theorem is false, and so there must be a counterexample, namely a map which cannot be coloured with four colours. Moreover, among all the counterexamples there must be one which is minimal, in the sense that it has the smallest possible number of countries. Then we try to prove all sorts of properties of a minimal counterexample, in the hope of getting a contradiction. If we succeed, then we have proved that a minimal counterexample cannot exist, and therefore no counterexample can exist, and therefore, the theorem is true in all cases.

In more colourful language, a minimal counterexample is sometimes called a 'minimal criminal'—if the 'law' (e.g. the four-colour conjecture) is broken, then there must be a criminal, and therefore, there must be a minimal criminal. If we hunt the minimal criminal, and find that he does not exist, then there can be no criminals, and the law is upheld.

Theorem 4.7 *Any minimal counterexample to the four-colour theorem contains at least 12 pentagons.*

Proof Kempe's argument shows that a minimal counterexample to the four-colour conjecture can have no triangles or quadrangles, and therefore, since by Proposition 3.15

$$\sum_{i=1}^{\infty} (6-i)r_i \geqslant 12,$$

has at least 12 pentagons. □

4.4 Vertex-colouring of graphs

Kempe's paper of 1879 already mentions the dual version of the four-colour problem, where we consider colouring the vertices of the dual pseudograph of a map. This version is much easier to formalize, and is the version which is almost always used nowadays. In particular, we do not need to worry about the faces, and such matters as whether a face meets itself at a vertex. Moreover, the vertex-colouring version lends itself more easily to generalization. Thus, the four-colour conjecture can be stated without any restriction on the pseudograph other than planarity. However, clearly we cannot allow loops, and multiple edges are irrelevant, so we might as well state the four-colour conjecture for graphs only.

Conjecture 4.8 *If G is any planar graph, then G is 4-vertex-colourable.*

The concept of a **Kempe chain** is now much easier to describe: it is a maximal connected subgraph consisting of vertices of two colours only. To illustrate this, we prove the dual version of Lemma 4.1.

Lemma 4.9 *Let G be a 4-vertex-colourable plane graph, and let a, b, c, d be the four vertices of a face of G, in cyclic order. Then G can be 4-coloured in such a way that a, b, c and d receive at most three colours between them.*

Proof If four distinct colours are used, then the graph cannot simultaneously contain a Kempe chain from a to c and a Kempe chain from b to d, for these two chains would have to intersect at a vertex, which is impossible since the colours of a and c are different from the colours of b and d (see Fig. 4.5). Therefore, we can change colours so that either a and c have the same colour, or b and d have the same colour. □

Remark This result can be generalized to the case where a, b, c, d is **any** cycle of length 4 in G (see Corollary 10.6).

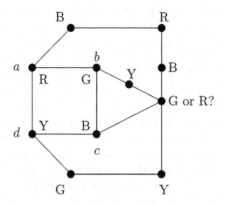

Fig. 4.5 Impossibility of two Kempe chains.

4.5   A three-colour theorem

We have now reduced the four-colour problem to the case of cubic maps. Whilst we cannot at present determine the precise number of colours required to colour the faces of all cubic maps, we can determine exactly which ones can be coloured with three colours.

First, we see that if any face of the map is surrounded by an odd number of edges, then the adjacent faces cannot be coloured with just two colours, since they would have to be coloured alternately (this uses the fact that the graph is cubic). Therefore, these faces require at least three colours, and the whole map requires at least four.

It follows that if a cubic map can be coloured with three colours, then every face is surrounded by an even number of edges. Another way of expressing this is to say that the vertices of the dual graph have even degree, which is exactly the criterion for the existence of an Euler circuit. Translating everything into the dual form, therefore, we have proved the following.

Theorem 4.10 *Let G be a plane graph in which all faces are triangles, and suppose that G is 3-vertex-colourable. Then, G has an Euler circuit.*

In fact, the converse is also true, as we proceed to prove.

Theorem 4.11 *Let G be a plane graph in which all faces are triangles, and suppose that G has an Euler circuit. Then G is 3-vertex-colourable.*

Proof   Let G be a counterexample with the minimum possible number of vertices. Since G has an Euler circuit, all vertex degrees are even, so the faces can be coloured with two colours by Exercise 2.4. We wish to show that there is a 3-colouring of the vertices in which the vertices of every white triangle are red, blue, green in clockwise order, while the vertices of every black triangle are red, blue, green in anticlockwise order.

Now by Corollary 3.19, there is a vertex of degree less than 6, and therefore, there is a vertex of degree 4. We start by colouring around a vertex of degree 4, as shown in Fig. 4.6. If the vertex itself is coloured green, then its neighbours are coloured alternately red and blue. Now extend the colouring to the neighbours of the two blue vertices: these neighbours are coloured alternately red and green. Thus, we obtain a ring of red and green vertices surrounding the two blue vertices and the central green vertex. Now collapse these three interior vertices to a single blue vertex v' joined to all the vertices in the red–green ring.

This gives us a new graph on fewer vertices, in which all faces are triangles. Moreover, the colouring of the triangles which remain is the same as the corresponding ones in the original graph. Therefore, by induction there is a 3-colouring of the vertices of this smaller graph. Without loss of generality, we may suppose the colours to be named so that v' is blue, and its neighbours are alternately red and green. Then the colouring can be extended to the original graph in the obvious way. This contradiction proves the theorem.   □

The following corollary is simply the dual version of this theorem, and was originally proved by Heawood in 1898.

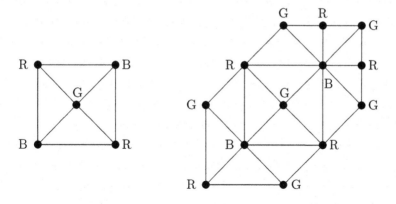

Fig. 4.6 Colouring around a vertex of degree 4.

Corollary 4.12 *A cubic map is 3-colourable if and only if every face has an even number of edges.*

Exercises

Exercise 4.1 Three-colour the faces of a cube and exhibit the three Kempe chains.

Exercise 4.2 Four-colour the faces of a dodecahedron and exhibit the six Kempe chains. Do you notice anything interesting?

Exercise 4.3 Suppose M is a 4-colourable map, and suppose that there is a vertex of M at which six countries meet. Show that if all four colours are used to colour these six countries, then the colours are used in one of the following five orderings (up to cyclic permutations, reflections, and renaming colours):

1. rbrgry
2. rbgrby
3. rbgryb
4. rbrbgy
5. rgrbyb

Exercise 4.4 Using Kempe chain arguments, show that the first four cases above can be reduced either to a colouring which uses only three colourings for the six countries, or to the fifth case.

Part II

Related topics

5

Other approaches to the four-colour problem

5.1 Hamilton cycles

There is a surprising connection between map-colouring and Hamilton cycles. We need to stay with the face-colouring version of the four-colour conjecture to make this connection explicit. A **Hamilton cycle** is a cycle which includes each vertex exactly once (in other words, it is a spanning cycle). A graph is called **Hamiltonian** if it has a Hamilton cycle. The origin of this term was Hamilton's 'icosian game' of 1857, whose object was to find such a cycle on the dodecahedron (see Fig. 5.1), although the general question of whether any given polyhedron possesses such a cycle was considered slightly earlier by Kirkman (see [31,32]). The idea of a Hamilton cycle can be traced back to 1759, when Euler introduced the knight's tour problem, though not of course in graph-theoretical language. In contrast to Euler circuits, there is no simple criterion to decide if a given graph has a Hamilton cycle. On the other hand, it turns out that if the underlying graph of a map has a Hamilton cycle, then the map is 4-colourable. This was apparently proved by Tait in 1880, as a special case of a more general result (see Theorems 5.7 and 5.8), although the published version of his lecture gives merely a sketch.

Theorem 5.1 *If a map has a Hamilton cycle, then it can be 4-coloured.*

Proof Consider the countries in the interior of the Hamilton cycle, and draw the part of the dual graph induced on them. This is a tree, for otherwise a cycle in the dual graph encloses a vertex of the original map (as in Fig. 2.11), which cannot, therefore, be visited by the Hamilton cycle (a contradiction). Now the vertices of a tree can always be 2-coloured, by working outwards from any given point. Therefore, the countries inside the Hamilton cycle can be coloured with two colours.

Similarly, the countries in the exterior of the cycle have the same property, so can be coloured with two other colours. Thus the whole map can be 4-coloured, as required.   □

For example, if we colour the interior faces of the Hamilton cycle in Fig. 5.1 alternately red and green, and the exterior faces alternately yellow and blue, we obtain the 4-colouring shown in Fig. 5.8(a).

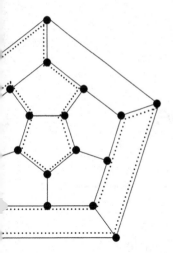

nilton cycle on the dodecahedron.

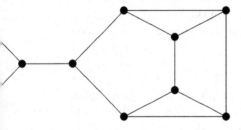

xample to a naive version of his conjecture.

ed the question as to whether all cubic graphs
liately rejects this, saying it is obvious the graph
e whose removal disconnects the graph). Clearly,
a Hamilton cycle, since no cycle can cross this
ve cannot use it again, and hence cannot get back
terexample is shown in Fig. 5.2. It seems that he
aphs he was considering were plane graphs. Next
mple (see Fig. 5.3), which in our terminology is
clude such cases, he imposes the extra condition
of a polyhedron', before dismissing the question

various generalizations of this conjecture were
[38] and Fig. 5.4), and König in 1936 (see [33]
ture was not disproved until 1946, when Tutte
e Fig. 5.6). It is quite complicated, and has 46
unterexample with 38 vertices and 21 countries

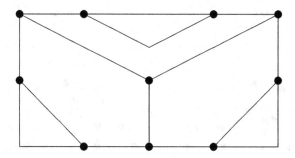

Fig. 5.3 Tait's multigraph counterexample to his conjecture.

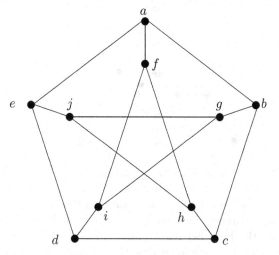

Fig. 5.4 The Petersen graph is not Hamiltonian.

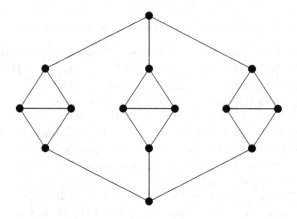

Fig. 5.5 König's bridgeless counterexample to Tait's conjecture.

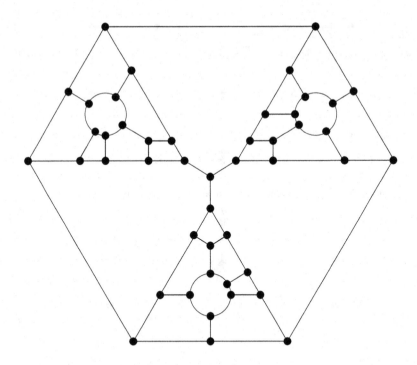

Fig. 5.6 Tutte's polyhedral counterexample to Tait's conjecture.

was constructed by Lederberg in 1967. According to Barnette [7], Okamura has shown that no counterexample exists with fewer than 34 vertices. If we drop the cubic condition, then the Herschel graph shown in Fig. 5.7 is a counterexample with 11 vertices. Moreover, it is known that there is no smaller counterexample to this version of the conjecture. We now consider some of these examples in detail.

Lemma 5.2 *The Petersen graph (Fig. 5.4) has no Hamilton cycle.*

Proof The Petersen graph (Fig. 5.4) consists of an outer pentagon $abcde$, an inner pentagon $fhjgi$, and five joining edges af, bg, ch, di, ej. A Hamilton cycle must enter the inner pentagon as often as it leaves it, and so contains an even number of the joining edges. If two, then they meet two adjacent vertices on either the outer or the inner pentagon, but not both, so we miss out at least one vertex on one of these pentagons. (In the figure, if the Hamilton cycle contains the edges af and bg, then on the inner cycle f and g are joined either by $fhjg$ or by fig, and in either case not all vertices of the inner cycle are on the Hamilton cycle. Similarly, if it contains the edges af and ch, then on the outer cycle a and c are joined either by abc or by $cdea$.) If four, it is again easy to get a contradiction: suppose the edges af, bg, ch, di are in the Hamilton cycle, so that ej is not, whence gj and hj are, in order to include j in the cycle. Similarly, in order to include e we must use edges ea and ed. But then the only way to

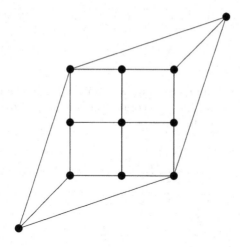

Fig. 5.7 The Herschel graph.

continue the paths is to add the edges bc and fi, which give two 5-cycles, not a 10-cycle. □

Since Tait's conjecture turns out to be false, we might consider trying to strengthen the polyhedral condition. By a theorem of Steinitz, this is equivalent to the condition that the graph is connected and planar, and cannot be disconnected by removing fewer than three vertices, that is, the graph is 3-connected (see Definition 7.4). Tutte [48] has shown that 4-connected planar graphs (with the obvious definition) always have Hamilton cycles. This is definitely not true for non-planar graphs, however. This is one of several known sufficient (but not necessary) conditions for a graph to be Hamiltonian. See the book by Bondy and Murty [14] for other examples.

The following result, due to Grinberg, is one of the few known simple necessary conditions for the existence of a Hamilton cycle. It can, therefore, be used to show that certain graphs do not have Hamilton cycles. However, it is no use for Tutte's graph, and in fact it is not easy to prove that Tutte's graph has no Hamilton cycle, except by trial and error.

Theorem 5.3 (Grinberg) *If G is a plane graph with a Hamilton cycle C, let r_i denote the number of i-sided faces inside C, and s_i the number of i-sided faces outside C. Then*

$$\sum_{i=2}^{\infty}(i-2)(r_i - s_i) = 0.$$

Proof The number of edges in C is p, the number of vertices of G, since C is a Hamilton cycle. Let e be the number of edges in the interior of C. Then the number of faces inside C is $e + 1$, since, as we have noted in the proof of Theorem 5.1, the subgraph of the dual graph induced on the corresponding vertices is a tree. But the total number of faces inside C is by definition $\sum_{i=2}^{\infty} r_i$, so we have

$$\sum_{i=2}^{\infty} r_i = e + 1.$$

Next, we count the total number of edges of all the faces inside C. On the one hand, this is $\sum_{i=2}^{\infty} i r_i$. But this just counts the edges of C once each, and the edges interior to C twice each. Thus,

$$\sum_{i=2}^{\infty} i r_i = 2e + p$$

$$\Rightarrow \sum_{i=2}^{\infty} (i-2) r_i = 2e + p - 2(e+1)$$

$$= p - 2.$$

Similarly, we obtain

$$\sum_{i=2}^{\infty} (i-2) s_i = p - 2,$$

and the result follows by subtraction.   □

Corollary 5.4 *The Herschel graph (Fig. 5.7) is not Hamiltonian.*

Proof   All nine faces of the graph are quadrangles, so with the notation of the theorem, we have $2(r_4 - s_4) = 0$, so $r_4 = s_4$ and the total number of faces is $r_4 + s_4 = 2r_4$, which is even. But this contradicts the fact that the number of faces is odd.   □

Alternatively, since the faces are quadrangles, it follows that the graph is bipartite. But a bipartite graph with a Hamilton cycle has both parts with the same number of vertices, and so has an even number of vertices altogether. This contradicts the fact that the Herschel graph has 11 vertices.

5.2   Edge-colourings

As we have seen, Tait observed in 1880 that if the underlying graph of a map has a Hamilton cycle, then the map can be 4-coloured. He was thus led to try (unsuccessfully, of course) to prove that every cubic map has a Hamilton cycle, and thereby prove the four-colour theorem. He also made the observation (Proposition 5.5) that, on a cubic map, the edges of a Hamilton cycle can be coloured alternately with two colours, and the remaining edge with a third colour. In this way, every vertex is incident with an edge of each colour. The relationship between such edge-3-colourings and 4-colourings of maps was explored first by Tait and later by Petersen.

We define an **edge-k-colouring** of a graph to be a colouring of the edges with k colours, in such a way that no two adjacent edges have the same colour. In other words, all the edges at each vertex must have different colours. It is obvious, therefore, that you need at least Δ colours, where $\Delta = \Delta(G)$ is the maximum degree of the vertices of the graph G. In some cases, this number of colours is sufficient, as in the following example.

Proposition 5.5 *Any Hamiltonian cubic graph can be edge-3-coloured.*

Proof Since G is cubic, we have $2q = 3p$, so p is even. Thus a Hamilton cycle has an even number of edges, so can be coloured with two colours. There is now just one more edge at each vertex, so these remaining edges cannot meet each other, which means they can all be coloured the same colour. □

In other cases, however, Δ colours may not be sufficient.

Proposition 5.6 *The Petersen graph is not edge-3-colourable.*

Proof If it is, supposing the colours are red, green and blue, then every vertex is adjacent to three edges, one of each colour. Therefore, the red and green edges together form a union of disjoint cycles of even length, which together cover all 10 vertices. Now the girth is 5, so there are no cycles of length 2 or 4, and the graph is not Hamiltonian, so there is no cycle of length 10. But there is no way to cover 10 vertices with disjoint cycles of lengths 6 and 8. □

In the above proof, we have used the fact that in any 3-colouring of the edges of a cubic graph, the edges of any two colours form a spanning set of disjoint cycles of even length. This connection was made explicit by Tait.

Theorem 5.7 (Tait) *Let G be a cubic graph (not necessarily planar). Then G can be edge-3-coloured if and only if G is spanned by a collection of disjoint cycles of even length.*

Proof If G is spanned by such a collection of cycles, then as above we can colour the edges of the cycles with two colours, and the remaining edges with a third colour. Conversely, if G is edge-3-coloured, then the edges of any given two colours form such a collection of cycles. □

The following important theorem describes the fundamental relationship between edge-colouring and the four-colour conjecture. Remember that we can assume that the underlying graph of our map is a bridgeless cubic plane graph. The theorem was originally stated by Tait [46], but his paper was only an abstract of his lecture, so contained only a sketch of the proof. A more complete proof was given by Petersen in 1898, when he also proved Proposition 5.6 and Theorem 5.7.

Theorem 5.8 (Tait) *Let G be a bridgeless cubic plane graph. Then the edges of G can be 3-coloured if and only if the faces of G can be 4-coloured.*

Proof Suppose first that we are given a 4-colouring of the faces. Then the edges are of six types, according to which pairs of colours lie on either side of it (Fig. 5.8). But a red–green edge cannot meet a blue–yellow edge, since two adjacent edges must be on the boundary of a face (since the graph is cubic), and therefore, share a colour. Thus, we can colour the red–green edges with the same

(a)

(b)

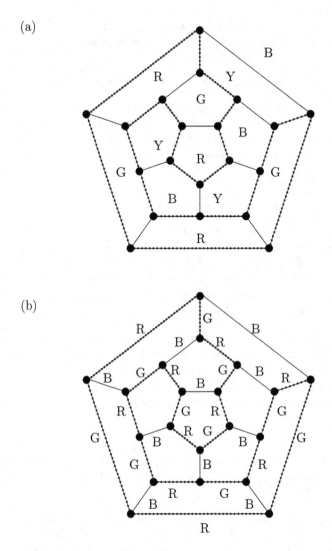

Fig. 5.8 Colouring the dodecahedron. (a) A 4-colouring of the faces and (b) a 3-colouring of the edges.

colour as the blue–yellow edges, and so on. Thus, we only need three colours for the six types of edges.

Conversely, suppose we are given a 3-colouring of the edges. Then the edges of any two colours (say red and green) form a collection of disjoint cycles, which together include all the vertices. Any face is inside some number of these cycles, and this number can be either even or odd. Now do the same for another pair of colours (say red and blue). We have now divided the faces into four types, which we assign to four colours. All we have to prove is that two adjacent faces cannot have the same colour. Now, two adjacent faces meet either at a red edge, or at a green edge, or at a blue edge. If they meet at a green edge, then crossing this edge changes the parity of the number of red–green cycles that a face is contained in. Similarly, if we cross a blue edge, then we change the parity of the number of red–blue cycles, while if we cross a red edge then we change both parities. □

Remark Note that if G has a bridge, then the exterior face lies on both sides of this bridge, so there is no 4-colouring of the faces according to the usual meaning. Similarly, a cubic graph with a bridge has an odd number of vertices on each side of the bridge, by the handshaking lemma, so cannot possibly be covered with even length cycles, and so, by Theorem 5.7, cannot be edge-3-coloured. This means that (technically speaking) we do not need the bridgeless condition in Theorem 5.8, since the theorem is vacuously true in the other cases.

By combining Theorems 5.7 and 5.8 we obtain Petersen's equivalent formulation of the four-colour conjecture. In a sense, this was already known to Tait, although he was at a disadvantage in believing the four-colour theorem to have been proved.

Theorem 5.9 *The four-colour conjecture is equivalent to the conjecture that every bridgeless cubic planar graph is spanned by a collection of disjoint cycles of even length.*

5.3 More on edge-colouring

We have just seen that 4-colouring of maps is equivalent to 3-colouring of cubic bridgeless (connected) plane graphs. It is obvious that cubic graphs need at least three colours for colouring the edges, and a simple induction shows that five colours are sufficient, since each edge is adjacent to just four others. We have seen that the Petersen graph requires four colours (though of course it is not planar). In fact, **every** cubic graph can be edge-coloured with four colours, and we will prove this in due course. This is a special case of Vizing's theorem, proved in 1964, and described by Fiorini and Wilson [21] as 'the great breakthrough'.

First, define the **chromatic index**, or **edge-chromatic number**, to be the minimum number of colours required to colour the edges of a graph, in such a way that adjacent edges have different colours. Thus, we require that all the edges incident with v have different colours. We write $\chi'(G)$ for the chromatic index. As above, it is obvious that $\chi'(G) \geqslant \Delta(G)$, and it is easy to see that $\chi'(G) \leqslant 2\Delta(G) - 1$ (just keep colouring the edges one at a time, and you will find you always have enough colours available for each edge, since no edge is adjacent to more than $2\Delta(G) - 2$ others).

The truly remarkable fact is that $\chi'(G) \leqslant \Delta(G) + 1$, so that $\chi'(G)$ is very tightly bounded. This is Vizing's theorem, which we prove in Theorem 5.11. It immediately suggests the question, for which graphs is $\chi'(G) = \Delta(G)$, and for which graphs is $\chi'(G) = \Delta(G)+1$? The answer to this question is far from known.

The proof we shall give of Vizing's theorem is by a Kempe chain argument. Note that Kempe chains for edge-colouring are very simple: they are either closed

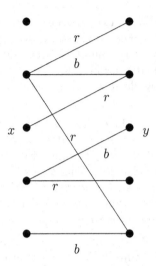

Fig. 5.9 The proof of König's theorem.

cycles of even length, or open paths. To illustrate the use of Kempe chains of edges, we first prove a much earlier result, due to König, which shows that bipartite graphs are of the first type, that is, have $\chi'(G) = \Delta(G)$. In fact König proved this theorem more generally, for multigraphs.

Theorem 5.10 (König, 1916) *If G is a bipartite multigraph, then $\chi'(G) = \Delta(G)$.*

Proof We use induction on the number of edges. The induction starts with a multigraph with no edges, in which case $\Delta(G) = 0$ and $\chi'(G) = 0$. Now suppose there is at least one edge. Then we remove an edge xy from G, to get a multigraph H, say (Fig. 5.9). Then we colour H with at most $\Delta(H)$ colours (and therefore with at most $\Delta(G)$ colours, since $\Delta(H) \leqslant \Delta(G)$). Now in H, both x and y have degree at most $\Delta(G) - 1$, since the edge xy has been removed, and therefore, there is at least one colour missing at x, and at least one colour missing at y. If the same one colour is missing at both x and y, then obviously we can colour xy with that colour, and the induction continues. So the only problem is if, say, x is incident with a red edge but no blue edge, and y is incident with a blue edge but no red edge. Now consider the red–blue Kempe chain starting from x. As we use red edges and blue edges alternately, we keep crossing from one of the two parts of the bipartite graph to the other, and back again. We cannot ever reach y, for each time we reach the y-part of the bipartite multigraph, we use a red edge. To put it another way, any path from x to y has an odd number of edges, so if the edges are alternately red and blue, the path starts and ends with the same colour.

Therefore, we can swap the colours red and blue in the Kempe chain from x, which leaves red available to colour xy. $\qquad\square$

Now we are ready to tackle the proof of Vizing's theorem. The argument we give closely follows that given by Fiorini and Wilson [21]. See also Bondy and Murty [14], who give a similar proof, which they attribute to Fournier.

Theorem 5.11 (Vizing, 1964) $\chi'(G) \leqslant \Delta(G) + 1$.

Proof As usual, we proceed by induction on the number of edges. Thus, we remove an edge vw_1, and colour the remaining edges with at most $\Delta(G) + 1$ colours. If there is a colour that is not used at v which is also not used at w_1, then we can use that colour for vw_1 and the induction continues.

Otherwise, we have a colour, red (r), say, used at w_1 but not v, and a colour, blue (b_1) say, used at v but not at w_1. As in the proof of König's theorem, if v and w_1 are not in the same red–blue Kempe chain, then we can recolour one of these two Kempe chains, to obtain a colour for vw_1, and the induction continues. So, we may assume that this does not happen, so v and w_1 are in the same red–blue $(r\text{–}b_1)$ Kempe chain (see Fig. 5.10(a)).

Now let vw_2 be the blue (b_1) edge from v. We want to use blue for the edge vw_1, so we do so, but then we have to look for another colour for the edge vw_2. Remember that w_1 and w_2 are still in the same red–blue Kempe chain (see Fig. 5.10(b)).

For the second step of the inductive process, we look at which colours are missing at v and at w_2. We still have a red edge used at w_2, and no red edge at v. Also, there is a blue (b_1) edge at v but not at w_2. Since at most $\Delta(G) - 1$ edges at w_2 are coloured, we may suppose that there is another colour, different from blue, say black (b_2), missing at w_2, but used at v. By the same process as before, we can suppose that v and w_2 are in the same red–black $(r\text{–}b_2)$ Kempe chain, and we let vw_3 be the black edge from v (see Fig. 5.10(c)). We then uncolour vw_3 so that we can colour vw_2 black (see Fig. 5.10(d)).

We keep on going in this way, until eventually the second colour which is missing at w_k is a colour b_j (say brown) which we have seen before (in other words, $j + 1 < k$). Now consider the red–brown $(r\text{–}b_j)$ Kempe chain containing v. By our assumptions, this chain runs from v through w_j to w_{j+1}, starting with a brown edge and ending with a red one (see Fig. 5.10(e)). Moreover, this chain does not contain w_k, since the colour brown is missing at w_k, whereas all vertices of the chain (except the two endpoints v and w_{j+1}) are incident with edges of both colours red (r) and brown (b_j). Therefore, we can change colour in the red–brown $(r\text{–}b_j)$ chain containing w_k without affecting the colours in the first chain. In particular, the missing colour at w_k changes from brown to red, which enables us to colour vw_k red (see Fig. 5.10(f)), and hence complete the colouring with $\Delta(G) + 1$ colours. $\qquad\square$

For examples, consider the complete graphs. If $G = K_{2n+1}$, then $\Delta(G) = 2n$, and the total number of edges is $(2n + 1)(2n)/2 = n(2n + 1)$. But at most n edges can be coloured with one colour, since each such edge uses up two vertices. Therefore there are at least $n(2n + 1)/n = 2n + 1 = \Delta(G) + 1$ colours.

An explicit colouring of K_{2n+1} with $2n+1$ colours may be obtained as follows. Draw the $2n + 1$ vertices equally spaced around a circle. Then each set of parallel

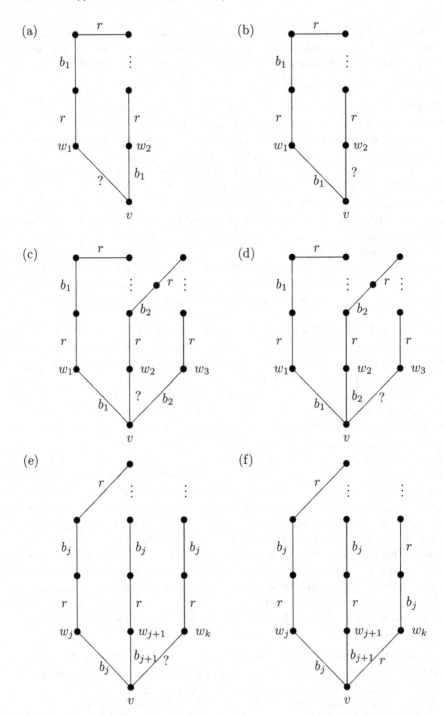

Fig. 5.10 The proof of Vizing's theorem.

lines contains n edges of the graph, and there are $2n+1$ such sets. We colour each set of parallel lines with one colour. The case $n = 2$ is illustrated in Fig. 5.11.

On the other hand, if $G = K_{2n}$, we first colour a subgraph K_{2n-1} with $2n-1$ colours as above. Then the colour missing at each vertex is different, so we can join each vertex to the last vertex with this missing colour, thereby achieving a colouring with $2n - 1 = \Delta(G)$ colours. The case $n = 3$ is illustrated in Fig. 5.12.

In fact, 'most' graphs can be edge-coloured with just $\Delta(G)$ colours (these graphs are called **class 1**), rather than needing $\Delta(G) + 1$ (these are called **class 2**). It turns out that there are just eight graphs on at most six vertices which are class 2. We have already seen two of them, namely K_3 and K_5. It is also easy to see that any odd cycle has class 2. The following proposition illustrates a useful method of proving that certain graphs are of class 2.

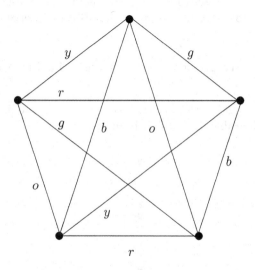

Fig. 5.11 A 5-colouring of the edges of K_5.

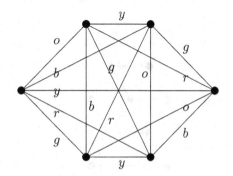

Fig. 5.12 A 5-colouring of the edges of K_6.

Proposition 5.12 *The graph G obtained by subdividing one edge of K_4 is of class 2.*

Proof The graph G is illustrated in Fig. 5.13. As there are five vertices, there can be at most two edges of any given colour. But there are seven edges altogether, so at least four colours are required. Thus $\Delta(G) = 3$ and $\chi'(G) = 4$, so G is of class 2. □

It has been conjectured that planar graphs are of class 1, provided $\Delta(G) \geqslant 6$. This has been proved by Vizing except in the cases $\Delta(G) = 6$ or 7.

There is a version of Vizing's theorem for multigraphs, in which 1 is replaced by the maximum number of parallel edges. That is, $\chi'(G) \leqslant \Delta(G) + m$, where m is the maximum number of parallel edges between any two vertices. Much more about edge-colourings can be found in the book by Fiorini and Wilson [21], including a great deal of discussion about distinguishing graphs of class 1 and class 2. See also the books by Bondy and Murty [14] and Berge [9].

Exercises

Exercise 5.1 Kirkman's graph is shown in Fig. 5.14. Prove that it is non-Hamiltonian.

Exercise 5.2 Show that Tutte's map (his counterexample to Tait's conjecture, see Fig. 5.6) is 4-colourable, but has no Hamilton cycle.

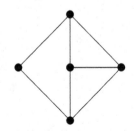

Fig. 5.13 A graph of class 2.

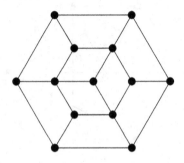

Fig. 5.14 Kirkman's graph of 1855.

Exercise 5.3 Find the chromatic index of the underlying graphs of each of the five Platonic solids.

Exercise 5.4 Do Proposition 5.5 and Theorems 5.7 and 5.8 remain true for multigraphs? Give proofs or counterexamples as appropriate.

Exercise 5.5 Find some graphs of class 2 other than those presented in this chapter.

Exercise 5.6 Prove that every cubic Hamiltonian graph is class 1.

Exercise 5.7 Find a planar graph G of class 2 with $\Delta(G) = 4$.

Exercise 5.8 Let G be the underlying graph of the icosahedron with one edge subdivided. Calculate p, q and r for this graph. Show that $\chi'(G) = \Delta(G) + 1$, so that G is of class 2.

Exercise 5.9 A famous problem known as Lucas's Schoolgirls Problem states: 'Each day $2n$ schoolgirls take a walk in pairs: for how many days can they walk before some pair walks together twice?' What is the answer, and why?

Exercise 5.10 Suppose that G is a graph satisfying $q > \Delta\lfloor\frac{1}{2}p\rfloor$. Prove that G is of class 2.

Show that this inequality can only hold if p is odd, and find an example of such a graph.

6

Maps on surfaces with holes

6.1 Some topology

It is a remarkable fact that although the map-colouring problem is so hard on the plane (or equivalently the sphere), it is much easier on the torus (a doughnut shape) or even on surfaces with larger numbers of holes. On the torus, for example, seven colours are necessary and sufficient. This was proved by Heawood in 1890, in the same paper in which he pointed out Kempe's mistake.

First, we need some background from topology. For the moment, we will only consider surfaces which, like the sphere, have an inside and an outside. Such surfaces are called **orientable**, as opposed to non-orientable surfaces such as the Klein bottle. A sphere is an example of an orientable surface which is **closed** (i.e. it has no boundaries), and **smooth** (i.e. it is infinitely differentiable, in a suitable sense). In fact, smooth closed orientable surfaces are characterized by their **genus**, which is just the number of 'holes' they have. Thus, the sphere has genus 0, since it has no holes, while a **torus** (the surface of a ring doughnut, or the inner tube of a bicycle tyre) has genus 1, since it has one hole through the middle.

An analogue to Euler's formula holds on a torus: $p - q + r = 0$. We have to be careful, though, to make sure that the faces are sensible: that is, we must be able to flatten out each face onto a plane without cutting it up. That means we cannot have a face so large that it contains the hole in the torus. In topological terms, each face must be **simply-connected**, or a **2-cell**, or **homeomorphic to a disk**. (These three terms are equivalent to each other in this context.)

As we have seen in Chapter 3, maps in the plane are equivalent to maps on a sphere. We can see this equivalence by using a projection from one to the other, or in purely topological terms we can imagine making a hole in the sphere and stretching it until it becomes flat. In a general sense this is how we put maps of a (nearly) spherical Earth into a flat Atlas.

For the sake of drawing clear pictures, we need to make similar flat maps representing real maps on curved surfaces of higher genus. If we want to flatten out a torus, we first of all need to cut a circle around the torus, and open it out into a cylinder. Then we need to cut open the cylinder along its length. We end up with a rectangle in which two opposite sides represent the same points, where the cyclinder was cut open. Similarly, the other two sides represent the two ends of the cylinder, which again came from the same points on the torus. The result

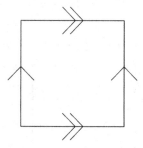

Fig. 6.1 Representation of a torus.

is as in Fig. 6.1, where the arrows are used to indicate that opposite edges are identified.

Reversing the whole process, a torus can be made from a (sufficiently flexible) square by first rolling it up into a cylinder, sticking two parallel edges together, and then rolling the cylinder up, sticking the two ends of the cylinder together. In other words, we identify the left and right edges of the square, and then identify the top and bottom edges (see Fig. 6.1).

Another way of picturing this is to repeat the square—as we cross the right-hand edge of the square, we reappear at the left-hand edge, so we can picture this by putting another copy of the square to the right of the first. If we keep doing this in all directions, we end up with a tiling of the plane with infinitely many identical squares (see Fig. 6.2).

Now imagine drawing maps on a torus. If there are no edges going through the hole, or round it, then we could remove the hole (either filling it in, in the first case, or breaking the ring, in the second case), and draw the map equally well on the surface of a sphere. Neglecting such trivial cases, we consider only the maps which really need the torus—these are called **2-cell embeddings**, which means that each region of the map can be flattened out into a disk without cutting the torus. In the other cases, there is some region which contains the hole, and cannot be flattened out in this way.

As in the plane, our map-colouring theorems rely on the corresponding version of Euler's formula for 2-cell embeddings of maps on a torus, that is $p - q + r = 0$.

Proposition 6.1 *If a connected graph G is drawn on a torus in such a way that every face is a 2-cell, and G has p vertices, q edges and r faces, then $p - q + r = 0$.*

Proof As in the proof of Theorem 3.2, we prove this by induction on the number of edges. If there is a vertex of degree 1, we remove it and the incident edge, without changing the value of $p - q + r$. If there is an edge which is incident with two different faces, remove it, thereby reducing q and r by 1 and not changing $p - q + r$. Moreover, the face formed by the union of the two old ones is still topologically a 2-cell, since it has no holes in it.

Otherwise, there is only one face, and all remaining edges are on the boundary of this face. By cutting along the boundary of this face, we can flatten out the

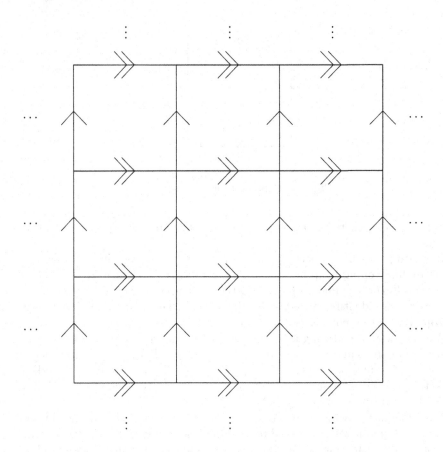

Fig. 6.2 Another representation of a torus.

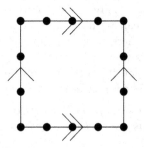

Fig. 6.3 A map with a single face on a torus.

torus, which means that it essentially looks as in Fig. 6.3. If there are m edges on each horizontal line in the diagram, and n edges in each vertical line, then there are $m + n$ edges in all, and $m + n - 1$ vertices. Since there is just one face, we have $p - q + r = (m + n - 1) - (m + n) + 1 = 0$. □

Corollary 6.2 *If G is any graph embedded in a torus, with p vertices, q edges and r faces, then*

$$p - q + r \geqslant 0.$$

Proof First, add just enough edges to make the graph connected. This can only reduce the value of $p - q + r$, as no new cycles are created, and therefore, the number of faces does not change. Now the new value of $p - q + r$ is either 2 or 0 according as the graph is planar or not, so the original value of $p - q + r$ was also at least 0. □

If we generalize to a surface with genus g, that is, one with g holes, then we obtain the following generalization of Euler's formula:

$$p - q + r = 2(1 - g).$$

Proposition 6.3 *If a connected graph G is drawn on a surface of genus g in such a way that every face is a 2-cell, and G has p vertices, q edges and r faces, then $p - q + r = 2 - 2g$.*

Proof (Sketch.) One way to prove this is to imagine boring a hole through the middle of one face, A, and reappearing in the middle of another face, B. We then join n of the vertices of A to n of the vertices of B, by n edges running through the hole. The result of this is to replace the two faces A and B by n faces, while adding n edges and leaving the number of vertices unchanged. Thus for each new hole, $p - q + r$ is decreased by 2, so the result follows by induction on the number of holes. □

Note that in this proof the induction starts with the sphere, so this gives us an alternative proof that $p - q + r = 0$ on the torus.

Corollary 6.4 *If G is any graph embedded in a surface of genus g, with p vertices, q edges and r faces, then*

$$p - q + r \geqslant 2 - 2g.$$

Proof First, add just enough edges to make the graph connected. This can only reduce the value of $p - q + r$, as no new cycles are created, and therefore, the number of faces does not change. Now the new value of $p - q + r$ is $2 - 2g'$, where $g' \leqslant g$ is the genus of the graph, so the original value of $p - q + r$ was also at least $2 - 2g'$, and therefore, at least $2 - 2g$. □

The quantity $2(1 - g)$ is called the **Euler characteristic** of the surface. One reason for using this concept rather than the genus is that it can be defined for non-orientable surfaces also (see Section 6.4).

6.2 Map colouring on a torus

Just as on the plane, k-colouring all maps on a torus is equivalent to k-colouring all cubic maps on the torus. The proof is just the same as the proof of Theorem 4.6. Corresponding to Corollary 3.13 we have the following.

Proposition 6.5 *In any cubic map drawn on a torus, there exists a face with at most six neighbours.*

Proof We have $2q = 3p$ for cubic maps, as before, and on the torus Euler's formula gives

$$0 \leqslant 3p - 3q + 3r = 3r - q$$

so $q \leqslant 3r$. If d denotes the average number of neighbours to a face, then by the handshaking lemma we have $2q = dr$ which implies that

$$d = \frac{2q}{r} \leqslant \frac{6r}{r} = 6.$$

Therefore, there exists a face with at most six neighbours. □

Note that if, in fact, all the faces are 2-cells, then the above argument shows that the average number of neighbours of the faces is exactly 6.

Theorem 6.6 *Every map on a torus can be coloured with at most seven colours.*

Proof If the map has seven or fewer faces, then the result holds trivially. If there are at least eight faces, then we can choose a face with at most six neighbours. Remove one edge from this face, and 7-colour the resulting smaller map—this is possible, by induction. Put the edge back, and colour the last face with a colour different from those of its (at most six) neighbours. □

This result is best possible, in the sense that there exists a map which requires seven colours. Indeed, there is a map consisting of seven hexagons, each of which is adjacent to all the other six (see Fig. 6.4 or Fig. 6.5). The dual graph of this map is, therefore, K_7, and is illustrated in Fig. 6.6.

Thus we have proved the following.

Theorem 6.7 (Heawood, 1890) *Seven colours are necessary and sufficient to colour all maps on a torus.*

Let us restate this in the dual form. We first define the **genus** of a graph to be the minimal genus of a surface in which it can be embedded without any edges crossing each other. Thus, planar graphs are exactly the graphs of genus 0. Given a map drawn on a surface of genus g in such a way that the individual countries are homeomorphic to disks, we define the underlying graph as before: we take vertices at the points where three or more countries meet, and edges for the portions of the boundaries running from one such point to another. The **dual graph** $D(M)$ of a map M is defined by taking a vertex $v(C)$ in $D(M)$ for each country C in M, and an edge joining $v(A)$ to $v(B)$ in $D(M)$ for each edge forming part of the boundary between A and B.

Now our original map-colouring problem translates to colouring the **vertices** of $D(M)$ in such a way that adjacent vertices receive distinct colours.

Theorem 6.8 *The vertices of any graph G of genus 1 can be coloured with at most seven colours.*

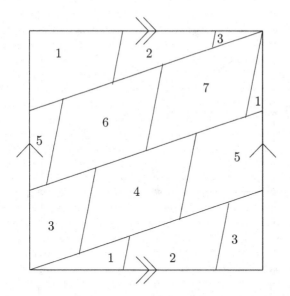

Fig. 6.4 A map on the torus requiring seven colours.

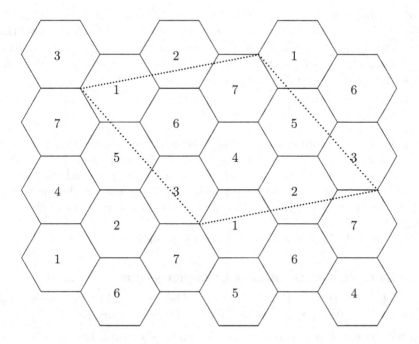

Fig. 6.5 Seven mutually adjacent hexagons on a torus.

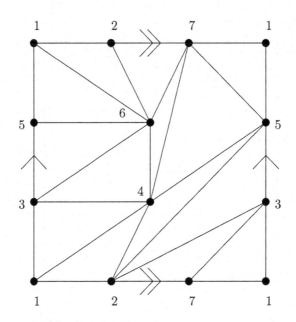

Fig. 6.6 K_7 embedded in the torus.

Proof By induction on the number of vertices. We may assume that G is connected. If G has at most seven vertices, then the result is trivial. So suppose that G has at least eight vertices. We may assume that the graph G has no multiple edges, so that the faces have at least three edges, so $3r \leqslant 2q$ by the handshaking lemma (the obvious generalization of Lemma 3.4). Therefore

$$0 = 3p - 3q + 3r \leqslant 3p - q$$

so $q \leqslant 3p$. But $2q = dp$ where d is the average degree of the vertices (by the other handshaking lemma, or Corollary 3.5), so $dp \leqslant 6p$, and $d \leqslant 6$. Thus, we can choose a vertex v of degree at most 6, and remove it and the incident edges. The resulting graph can be coloured with seven colours (by induction), and v has at most six neighbours, so can also be coloured with one of the seven colours.

Thus, the induction continues, and the result follows. □

6.3 Generalizing to surfaces of higher genus

In his 1890 paper Heawood proved not only Theorem 6.6, but also the generalization to $g > 1$. We first generalize Theorem 3.8 to arbitrary genus.

Proposition 6.9 *If a map is drawn on a surface of genus g, then*

$$q \leqslant 3p + 6g - 6.$$

Proof As usual, the faces have at least three sides, and so $2q \geqslant 3r$, and substituting into Euler's formula (Corollary 6.4) we have

$$6 - 6g \leqslant 3p - 3q + 3r \leqslant 3p - q,$$

and therefore,

$$q \leqslant 3p + 6g - 6$$

as required. $\qquad\square$

As an aside we remark that Theorem 3.9 can be generalized as follows.

Theorem 6.10 *If a graph of girth l is drawn on a surface of genus g, then*

$$q \leqslant \frac{l}{l-2}(p + 2g - 2),$$

or equivalently

$$g \geqslant 1 + \frac{1}{2}\left(\left(\frac{l-2}{l}\right)q - p\right).$$

Proof If the graph is disconnected, add just enough edges until it becomes connected, so that no new circuits are created. If the theorem is true for the new graph, then it is also true for the original graph. Since the new graph still has girth l, the handshaking lemma implies that $lr \leqslant 2q$. Moreover, the genus of the new graph is at most g, so substituting for r into $p - q + r \geqslant 2 - 2g$, we obtain

$$p - q + \frac{2q}{l} \geqslant p - q + r$$
$$\geqslant 2 - 2g$$
$$\Rightarrow p - q\frac{l-2}{l} \geqslant 2 - 2g$$
$$\Rightarrow q\frac{l-2}{l} \leqslant p + 2g - 2$$

which implies the first inequality by multiplying both sides by $l/(l-2)$, since $l \geqslant 3$. Rearranging again gives $2g \geqslant 2 - p + q(l-2)/l$, which implies the second inequality. $\qquad\square$

In particular, for a bipartite graph we have $g \geqslant 1 + \frac{1}{4}q - \frac{1}{2}p$, by putting $l = 4$ in the above theorem.

Returning to the proof of Heawood's theorem, we shall, as usual, consider the dual graph of a map. By the same construction as in the plane and on the torus, we can see that a map is k-colourable if and only if the dual graph is vertex-k-colourable. Also, without loss of generality we may restrict to the case when the vertices of the map all have degree 3. (If all such maps are k-colourable, then so are all maps.) This corresponds to the dual graph having all faces triangles.

Lemma 6.11 *If G is a graph drawn on a surface of genus g in such a way that all faces are triangles, then the average degree $d = d(G)$ of the vertices of G is*

$$d = 6 + \frac{12g - 12}{p},$$

where p is the number of vertices.

Proof Since all faces are triangles, we have $2q = 3r$, and therefore, from Euler's formula

$$6(1 - g) = 3p - 3q + 3r$$
$$= 3p - q$$

so $q = 3p + 6g - 6$. Moreover, $dp = 2q = 6p + 12g - 12$, so

$$d = 6 + \frac{12g - 12}{p}.$$

$\square$

Lemma 6.12 *With the same hypotheses, let $h = (7 + \sqrt{1 + 48g})/2$ be the positive root of the quadratic equation $x^2 - 7x + 12 - 12g = 0$. Then $d \leqslant h - 1$, provided $g \geqslant 1$.*

Proof The maximum degree of the vertices of G is at most $p - 1$, so $d \leqslant p - 1$, which implies

$$6 + \frac{12g - 12}{p} \leqslant p - 1,$$

and therefore, $12g - 12 \leqslant p^2 - 7p$, or

$$p^2 - 7p + 12 - 12g \geqslant 0.$$

Regarding this expression as a quadratic in p, the roots are

$$\frac{7 \pm \sqrt{1 + 48g}}{2}.$$

Only one of these roots is positive, so we deduce that

$$p \geqslant \frac{7 + \sqrt{1 + 48g}}{2} = h,$$

and h is a root of the above quadratic, so $h^2 - 7h + 12 - 12g = 0$, and therefore,

$$h - 7 = \frac{12g - 12}{h} \geqslant \frac{12g - 12}{p}.$$

Hence

$$d = 6 + \frac{12g - 12}{p}$$
$$\leqslant 6 + \frac{12g - 12}{h}$$
$$= 6 + (h - 7)$$
$$= h - 1$$

provided $g \geqslant 1$.

$\square$

Theorem 6.13 (Heawood, 1890) *If $g \geqslant 2$, then every graph on a surface of genus g is k-colourable, where*

$$k = \left\lfloor \frac{7 + \sqrt{1 + 48g}}{2} \right\rfloor .$$

Proof By induction on p. If $p \leqslant k$, the result is immediate. If $p > k$, then by Lemma 6.12, there exists a vertex v of degree at most $h - 1$. Of course, h need not be an integer, so we let $k = \lfloor h \rfloor$, the integral part of h, so that v has degree at most $k - 1$. Perform an 'elementary contraction' by identifying this vertex with any one of its neighbours. In other words, we replace the two vertices u, v by a single vertex x, joined to all the other vertices which were joined to either u or v, or both. The new graph has $p - 1$ vertices, so by induction can be vertex-coloured with k colours. Now we can k-colour the original graph, by colouring u the same colour as x, and colouring v (which has degree at most $k - 1$) with a colour not used by any of its neighbours. □

Definition 6.14 *The number*

$$k = \left\lfloor \frac{7 + \sqrt{1 + 48g}}{2} \right\rfloor$$

is called the **Heawood number** *of the surface.*

Heawood appeared to believe that he had also proved the converse of Theorem 6.13, but the fact that he did not was pointed out by Heffter. Indeed, this converse is much harder, and was not proved until 1968—we cannot prove it here. Clearly, it will be sufficient to embed the complete graph K_k in a surface of genus g, which will show that there is some map which requires exactly k colours. In 1957 Dirac proved that this condition is also necessary.

The cases $g \leqslant 6$ and some others were done by Heffter in 1891, but the general case was the work of many people, including Gustin, Terry, Welch, Guy and Mayer, culminating in the monumental work of Ringel and Youngs in 1968. The full proof is given in Ringel's book [40], and summarized in the article by White in [8, vol. 1, pp. 51–82]. This proof divides into 12 cases, according to the residue class of k modulo 12.

The following related result is an easy corollary of Theorem 6.13. For if K_n is k-colourable, then $n \leqslant k$. We give also a direct proof.

Proposition 6.15 *If the complete graph K_n on n vertices can be drawn as a map on a surface of genus g, then*

$$g \geqslant \frac{(n-3)(n-4)}{12},$$

or, equivalently

$$n \leqslant \frac{7 + \sqrt{1 + 48g}}{2}.$$

Proof   We have $p = n$ and $q = n(n-1)/2$, so substituting into the inequality $q \leqslant 3p + 6g - 6$ (see Proposition 6.9) gives

$$\frac{n(n-1)}{2} \leqslant 3n + 6g - 6,$$

which can be rearranged into the form

$$g \geqslant \frac{(n-3)(n-4)}{12}.$$

Now we can rearrange the inequality again to give $n^2 - 7n + 12 - 12g \leqslant 0$, and applying the usual criterion for a quadratic expression to be non-positive, we have

$$n \leqslant \frac{7 + \sqrt{1 + 48g}}{2}.$$

$\square$

As we have already noted, it is a fact, but very hard to prove, that the converse of this result is true. That is, the complete graph on n points can be drawn as a map on any surface whose genus is at least $(n-3)(n-4)/12$. Assuming this result, we can deduce the converse of Theorem 6.13.

Theorem 6.16   *If $g \geqslant 1$ then there is a map on the surface of genus g which requires k colours, where*

$$k = \left\lfloor \frac{7 + \sqrt{1 + 48g}}{2} \right\rfloor.$$

Corollary 6.17   *On a surface of genus g, where $g \geqslant 1$,*

$$k = \left\lfloor \frac{7 + \sqrt{1 + 48g}}{2} \right\rfloor$$

colours are both necessary and sufficient to colour all maps.

6.4   Non-orientable surfaces

It turns out that in addition to the orientable surfaces described above, which are classified by their genus g, or alternatively by their **Euler characteristic** $2 - 2g$, there is also a series of non-orientable surfaces. These may be defined by cutting n circular holes in a sphere, and sewing up each hole by identifying each point on a circle with the opposite point. This cannot be physically done in three dimensions, so is hard to visualize. These non-orientable surfaces do not have a genus, but they do have an **Euler characteristic**, which is $2 - n$. The case $n = 1$ is the **projective plane** (see Exercise 6.8), and the case $n = 2$ is the **Klein bottle** (see Exercise 6.10).

Many of the above results generalize to non-orientable surfaces, by way of the Euler formula $p - q + r = 2 - n$. In particular, Heawood's theorem applies to non-orientable surfaces, if we replace $2g$ by n in the formula, provided $n \geqslant 2$. Thus, the projective plane is the only case where it does not hold.

Theorem 6.18 *If* $n \geqslant 2$, *then every map on a non-orientable surface with Euler characteristic* $2 - n$ *is* k-*colourable, where*

$$k = \left\lfloor \frac{7 + \sqrt{1 + 24n}}{2} \right\rfloor .$$

On the other hand, the question of whether there **exists** a graph which requires k colours surprisingly has a different answer for the Klein bottle—in fact only six colours are necessary in this case, not the seven colours suggested by Heawood's formula (see Exercise 6.10).

Exercises

Exercise 6.1 Show how to draw the complete bipartite graph $K_{3,3}$ on a torus. Verify that Euler's formula holds for this embedding of the graph. How many colours are needed to colour the faces? How many colours are needed to colour the vertices?

Exercise 6.2 Do the same for the Petersen graph.

Exercise 6.3 Show that K_5 can be drawn on the torus as a 'regular' polyhedron with five faces, each with four sides.
  Show also that in this embedding, K_5 is self-dual.
  Can you find another embedding of K_5 on the torus, in which it is not self-dual?

Exercise 6.4 Suppose that the complete graph K_n can be drawn on a surface of genus g in such a way that all the faces are triangles. Use Euler's formula to prove that

$$12(g - 1) = n^2 - 7n.$$

Deduce that $n \equiv 0, 3, 4$ or $7 \pmod{12}$.

Exercise 6.5 Suppose that the complete bipartite graph $K_{m,n}$ can be drawn on a surface of genus g in such a way that all the faces are quadrangles. Use Euler's formula to prove that

$$(m - 2)(n - 2) = 4g.$$

Find all solutions to this equation in the case $g = 1$, and in the case $g = 2$.
  For each solution, determine whether or not the corresponding $K_{m,n}$ can be drawn on a torus. Hence, list all the complete bipartite graphs which have genus 1.

Exercise 6.6 Show that if $h = (7 + \sqrt{1 + 48g})/2$ is an integer, then

$$h \equiv 0, 3, 4 \text{ or } 7 \pmod{12}.$$

Calculate the values of g for the first few integer values of h.

Exercise 6.7 A surface of genus 2 can be represented in the plane by drawing an octagon, and identifying each edge with the opposite edge, as shown in Fig. 6.7.

Heawood's formula (Theorem 6.16) shows that (how many?) colours are sufficient to colour any map on this surface. Can you find an embedding of the complete graph on this number of points, into the surface, and thus show that this number of colours is also necessary?

Exercise 6.8 The **projective plane** is a surface which can be defined by identifying opposite sides of a square in **opposite** directions as in Fig. 6.8.

Show how to draw K_6 on a projective plane. Is your embedding a 2-cell embedding?

Calculate $p - q + r$ for this embedding. What does this tell you about the surface?

Prove the six-colour theorem for the projective plane.

Exercise 6.9 A **Klein bottle** is a closed non-orientable surface which may be represented by the diagram in Fig. 6.9. In other words, first make a cylinder, and then turn one end inside out (you can only 'physically' do this in a space of four or more dimensions), before sticking the two ends together.

Show that the analogue of Euler's formula on the Klein bottle is $p - q + r = 0$, just as for the torus.

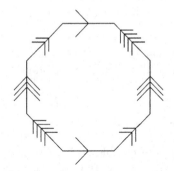

Fig. 6.7 Representation of a surface of genus 2.

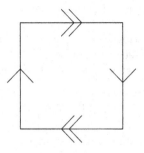

Fig. 6.8 Representation of a projective plane.

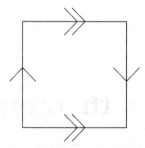

Fig. 6.9 Representation of a Klein bottle.

Exercise 6.10 (Hard) Use a Kempe chain argument to show that if six regions meet at a point in a 6-colourable map drawn on a Klein bottle, then the map can be 6-coloured in such a way that at most five colours are used for the given six regions.

Deduce that any map on a Klein bottle can be coloured with at most six colours.

7
Kuratowski's theorem

7.1 Connectivity

We have looked at problems of colouring graphs of different genus, but how do we know what genus a given graph has? In general, this is a hard problem. The genus 0 case is obviously of special interest to us, as a good characterization of genus 0 (i.e. planar) graphs might help us to attack the four-colour theorem. The most famous characterization of planar graphs is Kuratowski's theorem, which gives us a practical way of showing that a given graph is non-planar. On the other hand, the easiest way to show that a given graph is planar is to draw a plane graph which is isomorphic to it.

As we have already seen in Chapter 3, the complete graph on five vertices, K_5, is non-planar, as is the complete bipartite graph $K_{3,3}$. What Kuratowski's theorem tells us is that, in a certain sense, every non-planar graph 'contains' one of these two graphs.

In order to state the theorem precisely, we define a **subdivision** of a graph G to be a graph obtained from G by a finite number of operations of the following form: introduce a new vertex x and replace an edge vw by two edges vx and xw. We think of this as just adding a vertex in the middle of an existing edge. Obviously, if a graph is planar then any subdivision of it will be planar, and vice versa, so any subdivision of a non-planar graph is itself non-planar. In particular, since K_5 and $K_{3,3}$ are non-planar, it follows that any graph which contains a subdivision of either K_5 or $K_{3,3}$ is non-planar (Fig. 7.1). What is remarkable is that this is a necessary condition for a graph to be non-planar, as well as a sufficient one.

Theorem 7.1 (Kuratowski) *If G is a non-planar graph, then it contains a subgraph H which is a subdivision of K_5 or $K_{3,3}$.*

There is a similar theorem, due to Wagner, which uses **contractions** instead of subdivisions. An **elementary contraction** of a graph is the operation of replacing two adjacent vertices by a single vertex: the new vertex is joined to every other vertex which was joined to one or both of the original two vertices. A **contraction** of G is, then, any graph which can be obtained from G by a finite sequence of elementary contractions.

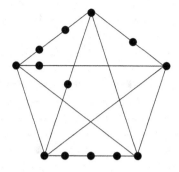

Fig. 7.1 A subdivision of K_5.

Theorem 7.2 (Wagner) *If G is a non-planar graph, then it contains a subgraph H which has K_5 or $K_{3,3}$ as a contraction.*

We prove Kuratowski's theorem first. The proof is quite hard: the version we give is a mixture of those given by Harary [28] and by Bondy and Murty [14]. Before we embark on it, we make a few useful definitions.

Definition 7.3 *The* **connectivity** *$\kappa(G)$ of a graph G is the minimum number of vertices you need to remove in order to disconnect the graph (or to reduce it to a 1-vertex graph, in the case when G cannot be disconnected by removing vertices).*

In this definition, note that when you remove a vertex you must also remove all the edges incident to that vertex, since an edge cannot live without its endpoints. We shall see in Exercise 7.6 that the complete graphs are the only connected graphs which cannot be disconnected by removing vertices.

Definition 7.4 *A graph is called k-**connected** if $k \leqslant \kappa(G)$, that is if it requires the removal of at least k vertices (and the incident edges) to disconnect the graph, or to make it a 1-vertex graph.*

In particular, a 1-connected graph is the same thing as a connected graph. A maximal 1-connected subgraph of a graph G is a connected component of G. Note that our definition implies that every k-connected graph has at least $k + 1$ vertices. Also note that if G is k-connected, then G is l-connected for all $l \leqslant k$. The complete graphs K_n are $(n - 1)$-connected, but not n-connected.

More generally, if G has a vertex v of degree d, then removal of all the neighbours of v will disconnect the graph, so G is at most d-connected. In other words, $\kappa(G) \leqslant \delta(G)$. For example, a tree is 1-connected but not 2-connected. A cycle is 2-connected but not 3-connected. The octahedron (Fig. 3.1(c)) is 4-connected but not 5-connected. Figure 7.2 shows the cycle C_6, which has connectivity 2, and the wheel W_7, which has connectivity 3.

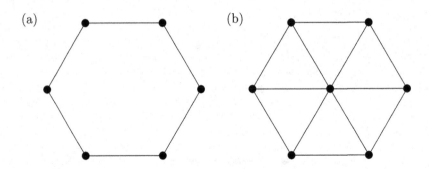

Fig. 7.2 Two-connected and 3-connected graphs. (a) $\kappa(C_6) = 2$ and (b) $\kappa(W_7) = 3$.

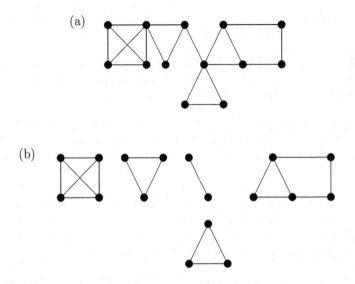

Fig. 7.3 The blocks of a 1-connected graph. (a) A graph with connectivity 1 and (b) its five blocks.

Recall that a **cutvertex** (sometimes called a **cutpoint** or **articulation vertex**) of a graph is a vertex whose removal disconnects the graph. Thus, a connected graph with a cutvertex is 1-connected but not 2-connected. Recall also that a **bridge** is an edge whose removal disconnects the graph.

Definition 7.5 *Let G be a graph. Then, the **blocks** of G are (a) the maximal 2-connected subgraphs and (b) the bridges with their two endpoints.*

In Fig. 7.3, we give an example of a 1-connected graph and its blocks.

Definition 7.6 *A graph G is k-**edge-connected**, if it requires the removal of at least k edges to disconnect the graph. The **edge-connectivity** $\kappa'(G)$ of a graph*

G is the minimum number of edges you need to remove in order to disconnect the graph. Thus, *G* is *k*-edge-connected if and only if $k \leqslant \kappa'(G)$.

These numbers are related by the following easy inequalities.

Theorem 7.7 $\kappa(G) \leqslant \kappa'(G) \leqslant \delta(G)$, where $\delta(G)$ is the minimum degree of the vertices of *G*.

Proof To prove the second inequality, choose a vertex *v* of minimal degree, that is of degree $\delta(G)$, and remove all the edges incident with *v*. This disconnects *G* by removing $\delta(G)$ edges, so $\kappa'(G) \leqslant \delta(G)$.

To prove the first inequality, choose a disconnecting set of $\kappa'(G)$ edges. Now remove one of the two vertices incident with each such edge. This forces us to remove the edges also, and therefore, disconnects the graph, removing at most $\kappa'(G)$ vertices. Therefore, $\kappa(G) \leqslant \kappa'(G)$. □

The next lemma is crucial to our proof of Kuratowski's theorem. It may be thought of as a special case of Menger's theorem, which states that if *G* has at least $k + 1$ vertices, then *G* is *k*-connected if and only if every pair of vertices *u*, *v* is connected by at least *k* internally disjoint paths (i.e. paths which intersect only at *u* and *v*). The 'if' part of Menger's theorem is easy: if *u* is connected to *v* by at least *k* internally disjoint paths, then you need to remove at least one internal vertex from each of these paths in order to disconnect *u* from *v*. The lemma we require, however, is the case $k = 2$ of the 'only if' part. This states that if *G* is 2-connected, then every pair of vertices is connected by two internally disjoint paths. In other words,

Lemma 7.8 If *G* is a 2-connected graph, then every pair of vertices lies on a cycle.

Proof Let *u*, *v* be distinct vertices, and suppose that there is no cycle containing both *u* and *v*. We aim for a contradiction. Choose a point *w* which is on a cycle with *u*, and is as close to *v* as possible, in the sense that there is a path from *w* to *v* with as few edges as possible. Note that there is some non-trivial cycle involving *u*, for otherwise one of the edges incident with *u* is a bridge, contradicting 2-connectedness. In particular, $w \neq u$. Now choose a cycle *u*–*w*–*u* and a shortest path from *w* to *v* (see Fig. 7.4).

Now 2-connectedness implies that there is a path *P* from *u* to *v* which does not pass through *w*. This new path *P* involves at least one of the vertices in the

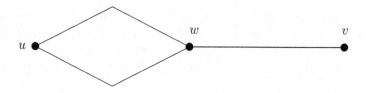

Fig. 7.4 No cycle through *u* and *v*.

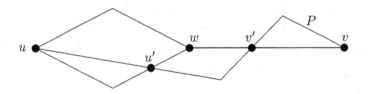

Fig. 7.5 Extending the cycle to v'.

u–w–u cycle (in particular, it involves u), so let u' be the last such vertex along P in the direction from u to v. Similarly, the section of the new path P from u' to v involves at least one vertex in the old w–v path (in particular, it involves v), so let v' be the first such vertex (see Fig. 7.5).

Then, the new path from u' to v' is internally disjoint from all the old paths, and we can see in the picture that there is a cycle through u and v', taking the old path from v' to w, then the old cycle from w to u and on to u', and finally, the new path from u' to v'. This contradicts our choice of w. □

7.2   A minimal counterexample to Kuratowski's theorem

We shall prove Kuratowski's theorem by contradiction, so suppose the theorem is false, which means there is a non-planar graph which does not contain a subdivision of either K_5 or $K_{3,3}$. Let G be such a graph with the minimum possible number of edges. (Such a graph is a **minimal counterexample** to Kuratowski's theorem.)

Lemma 7.9 *G is 3-connected.*

Proof Suppose G is not 3-connected, and choose vertices u and v which disconnect G. Thus, G is the union of two graphs which intersect in u, v, and possibly the edge uv. Now add the edge uv to each of these subgraphs if it is not there already. Then either both the resulting graphs are planar, or at least one of them is not planar.

In the latter case, we have constructed a smaller counterexample to Kuratowski's theorem, which is a contradiction. In the former case, we can draw each of the two graphs in the plane, with the edge uv on the exterior face. Then, we can join them together along this edge, and produce a planar embedding of G, which is again a contradiction. □

Now we choose any edge u_0v_0 of G, and let F be the graph obtained by removing that edge. By the minimality of G, we know that F is planar. For if F is not planar, then by assumption it is not a counterexample to Kuratowski's theorem, so it contains a subgraph which is a subdivision of K_5 or $K_{3,3}$, and therefore, so does G.

Corollary 7.10 *F is 2-connected.*

Proof If F can be disconnected by removing a vertex x, then G can be disconnected by removing u_0 (and hence the edge u_0v_0) and x. This contradicts Lemma 7.9. □

In particular, we have the following.

Lemma 7.11 *The graph F contains a cycle C going through both u_0 and v_0.*

Proof This follows immediately from Corollary 7.10 and Lemma 7.8. □

7.3 The proof of Kuratowski's theorem

With G, F and C as in the previous section, we choose the cycle C and a planar embedding of F, such that C has as many faces inside it as possible. Now F must be such that it is impossible to draw the edge u_0v_0 without crossing some edges of F. That means that u_0 and v_0 must be separated by both a piece of the graph inside C, and by a piece of the graph outside C.

To make these notions precise, consider the 'outer' subgraph F_1 consisting of all vertices and edges outside C, together with those vertices of C which are incident with such edges. We define an 'outer piece' to be a connected component of F_1. For convenience, we divide the cycle C into a left-hand path C_l running anticlockwise (say) from u_0 to v_0, and a right-hand path C_r running clockwise from u_0 to v_0 (see Fig. 7.6).

Lemma 7.12 *Each outer piece consists of a single edge with its two endpoints, one of which is in C_l and the other in C_r.*

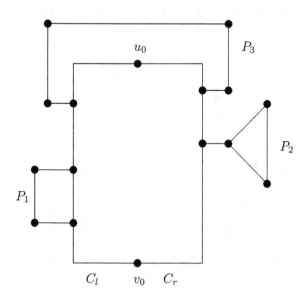

Fig. 7.6 Starting the proof of Kuratowski's theorem.

Proof First note that an outer piece P cannot meet C_l in more than one vertex, for if it did we could change C to enclose more faces, by taking a detour through P (see the piece P_1 in Fig. 7.6). Similarly, P cannot meet C_r in more than one vertex. Therefore, P meets each of C_l and C_r in exactly one vertex, since otherwise P meets C in only one vertex, and then that vertex is a cutvertex of F, contradicting the fact that F is 2-connected (see the piece P_2 in Fig. 7.6). Indeed, P must be a single edge, since otherwise its two endpoints in C not only disconnect F, but also disconnect G, contradicting the fact that G is 3-connected (see the piece P_3 in Fig. 7.6). □

As a consequence of this lemma, we can order the outer pieces vertically, so that P_i meets C_l in u_i, and meets C_r in v_i, in such a way that C_l passes through $u_0, u_1, u_2, \ldots, u_n, v_0$ in turn, and C_r passes through $u_0, v_1, v_2, \ldots, v_n, v_0$ in turn (see Fig. 7.7).

Next, we turn our attention to the inner pieces of F, defined analogously. Those which do not separate **any** u_i from the corresponding v_i can be redrawn outside C, since they do not cross any of the outer pieces P_i. If, having done this, none of the remaining inner pieces separates u_0 from v_0, then we can draw the edge $u_0 v_0$ inside C without violating planarity (see Fig. 7.8). This is a contradiction. Therefore, at least one of the inner pieces separates u_0 from v_0, and simultaneously separates some u_i from v_i. For the sake of the rest of the argument, we might as well suppose that $i = 1$.

We now disjoin cases according to where this inner piece meets the cycle C. Let w_0 and x_0 be places where it separates u_0 from v_0, and let w_1 and x_1 be places where it separates u_1 from v_1. For definiteness, say w_1 is between u_1 and v_1 going anticlockwise around C, and x_1 is between v_1 and u_1.

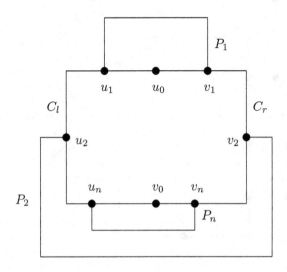

Fig. 7.7 Some outer pieces of F.

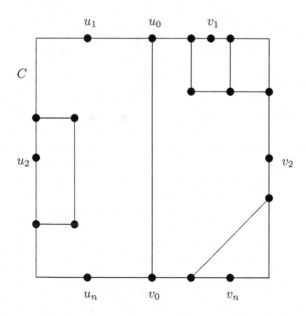

Fig. 7.8 Drawing u_0v_0 inside C.

Note that there is a symmetry of the whole situation interchanging the subscripts 0 and 1. For if we replace the edge u_0v_0 and draw it outside the cycle C, then the problem is to prevent the drawing of the edge u_1v_1 inside C. This symmetry can be used to reduce the number of cases which need to be considered.

Case 1. w_1 and x_1 are on opposite sides of u_0v_0.
Case 2. w_1 and x_1 are on the same side of u_0v_0.
Case 3. One of these points is equal to u_0 or v_0.
Case 4. Both of these points are equal to u_0 or v_0.

In each case, we throw away as much of the graph as we like, and ignore subdivisions of edges, until we obtain either K_5 or $K_{3,3}$. This will show that in every case we have a subgraph which is a subdivision of K_5 or $K_{3,3}$ (see Figs 7.9 and 7.10).

Case 1. In this case we get $K_{3,3}$ (Fig. 7.9(a)).
Case 2. In this case, the inner piece must also separate u_0 and v_0 on the other side. Either it meets the cycle at say v_1, or it meets at some other point. In the first case (Fig. 7.9(b)), we omit the path v_1v_0 and obtain a subdivision of $K_{3,3}$. In the second case (Fig. 7.9(c)), there is a subgraph looking just like case 1.
Case 3. Without loss of generality we have $x_1 = u_0$, and w_1 is below u_1 and v_1 in the cycle, so without loss of generality w_1 is between v_0 and v_1. In order to separate u_0 from v_0, this piece meets C also in C_l, either above, below or at u_1. The first case (Fig. 7.9(d)) contains case 1 as a subgraph, while the

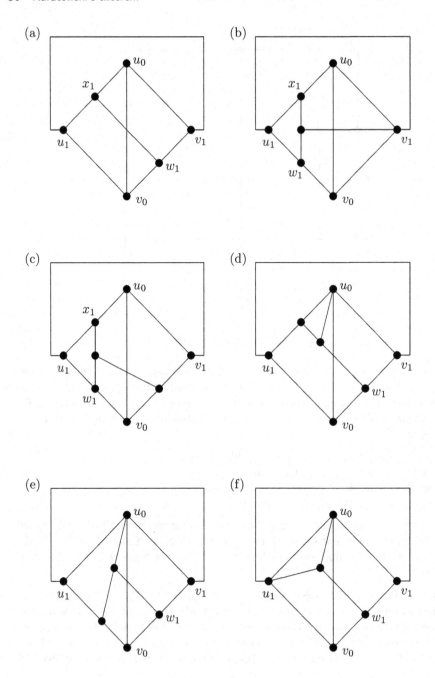

Fig. 7.9 Cases 1–3 in the proof of Kuratowski's theorem.

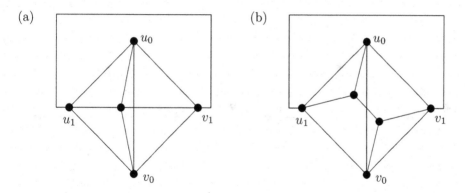

Fig. 7.10 Case 4 in the proof of Kuratowski's theorem.

second (Fig. 7.9(e)) is equivalent to case 2 on interchanging u_0 and v_0 with u_1 and v_1. The third case (Fig. 7.9(f)) gives a subdivision of $K_{3,3}$ on removing the path u_0u_1.

Case 4. By symmetry (i.e. by interchanging u_0 with u_1 and v_0 with v_1 if necessary) we can deal with all the cases except those where both w_0 and x_0 are equal to u_1 or v_1. There are now two cases: either the paths w_1x_1 and w_0x_0 meet at a single point, which gives us a subdivision of K_5 (Fig. 7.10(a)), or they overlap in a non-trivial path, in which case we have a subdivision of $K_{3,3}$ (Fig. 7.10(b)).

This concludes the proof of Kuratowski's theorem.

Let us now consider Wagner's theorem (Theorem 7.2). Historically, this was first proved as a corollary to Kuratowski's theorem, as here, but in the next section we shall give a direct proof.

Before we prove the theorem, note that if G is planar, and G' is the graph obtained from G by contracting the edge vw, then G' can be drawn in the plane by physically shrinking the edge vw until v and w coincide. Therefore, G' is also planar. It follows, by induction, that any contraction of a planar graph is planar.

Theorem 7.13 *A graph G is planar if and only if it contains no subgraph which has K_5 or $K_{3,3}$ as a contraction.*

Proof If G is not planar, then by Kuratowski's theorem (Theorem 7.1) G contains a subgraph H which is a subdivision of K, where K is either K_5 or $K_{3,3}$. Then H can be contracted onto K simply by contracting the subdividing edges one at a time. Thus, G contains a subgraph H which has K_5 or $K_{3,3}$ as a contraction.

Conversely, if H contracts onto K and K is not planar, then H cannot be planar, for if it were, then K would be planar by the above remarks. □

For example, consider the Petersen graph. If you contract the five edges connecting the outer 5-cycle with the inner one, then you obtain K_5 (see Fig. 7.11). Hence the Petersen graph is non-planar.

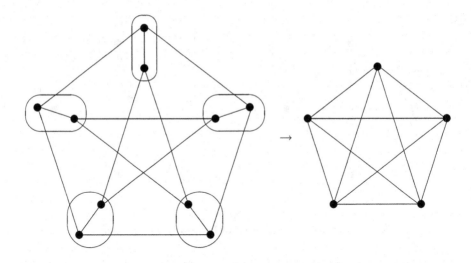

Fig. 7.11 Contracting the Petersen graph onto K_5.

Remark Notice that we have used the fact that if G is a subdivision of H, then G contracts onto H. It is worth noting that the converse of this is false (see the next section).

7.4 An alternative approach

Although Kuratowski's theorem is older than Wagner's theorem, and historically, the latter was deduced from the former, it is perhaps easier to prove Wagner's theorem first, and deduce Kuratowski's theorem from it. This is the approach adopted by Diestel [19], whose proof goes roughly as follows.

We take G to be a minimal counterexample to Wagner's theorem. Thus, G is a minimal non-planar graph which has no subgraph contracting onto K_5 or $K_{3,3}$. In particular, if we contract any edge of G, then we obtain a planar graph, since we obviously cannot have a subgraph contracting onto K_5 or $K_{3,3}$.

We first show that G is 3-connected, in exactly the same way as in Section 7.2.

Lemma 7.14 *Any minimal counterexample to Wagner's theorem is 3-connected.*

Proof Just as in Lemma 7.9. $\qquad\qquad\square$

Next we show that we can choose an edge to contract, in such a way that the resulting graph is still 3-connected.

Lemma 7.15 *If G is any 3-connected graph other than K_4, then G has an edge such that the graph resulting from G by contracting that edge is still 3-connected.*

Proof If not, then contracting **any** edge in G results in a non-3-connected graph. So if uv is any edge in G, then there is a vertex x such that $\{u, v, x\}$ disconnects G, into components $G_1, G_2, \ldots$, say. Moreover, each of u, v, x is adjacent to some vertex in each component G_i, for otherwise a proper subset of $\{u, v, x\}$ would already disconnect G.

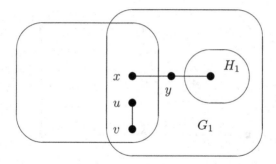

Fig. 7.12 The proof of Lemma 7.15.

Now choose uv and x so that there is a component G_1 with the smallest possible number of vertices, and choose y adjacent to x in G_1. Applying the same argument now to the edge xy, we find a vertex z such that $\{x, y, z\}$ disconnects G, into components $H_1, H_2, \ldots$, say. We shall show that one of these components has fewer vertices than G_1, and this contradiction will prove the lemma (see Fig. 7.12).

Let H_1 be a component not containing u, and therefore, not containing v (since u and v are adjacent). Now H_1 contains at least one neighbour of y, but does not contain u, v or x, so is entirely contained within G_1. On the other hand, H_1 does not contain y, so H_1 is a subgraph of G_1, with strictly fewer vertices, as required. $\square$

Theorem 7.16 (Wagner) *If G is a non-planar graph, then it contains a subgraph H which has K_5 or $K_{3,3}$ as a contraction.*

Proof We let G be a minimal counterexample, so that, by Lemma 7.14, G is 3-connected, and certainly G is not K_4, so by Lemma 7.15 we can choose an edge uv in G such that contracting uv gives a 3-connected graph F. By minimality of G, we know that F is planar, so we draw it in the plane, with a vertex w corresponding to the two vertices u and v in G. Then the faces incident with w together form a region R of the plane, enclosing w, and with a cycle C as its boundary. We now try to draw u and v inside R, and join them up to the appropriate vertices of the cycle.

First draw u and join it to its neighbours in C—let these be $u_1, u_2, \ldots, u_k$, $u_{k+1} = u_1$ in cyclic order around C. We now disjoin cases according to where in the cycle the neighbours of v lie. Notice that v has at least two neighbours v_1 and v_2 on the cycle, since its only other neighbour is u, and G is 3-connected. If there are neighbours v_1 strictly between u_i and u_{i+1} and v_2 between u_j and u_{j+1}, with $i \neq j$, then contracting the edges of the cycle between u_{i+1} and u_j if necessary, we obtain a $K_{3,3}$ on the vertex set $\{u, v_1, v_2\} \cup \{v, u_i, u_{i+1}\}$ (see Fig. 7.13(a)). The same thing happens if v_1 is as above and $v_2 = u_j$, with $j \neq i, i+1$ (see Fig. 7.13(b)).

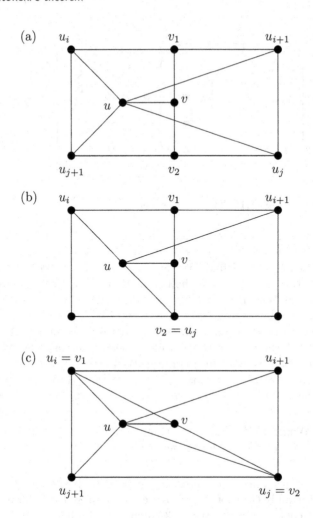

Fig. 7.13 Three cases in the proof of Wagner's theorem.

Otherwise, all neighbours of v in the cycle are neighbours of u. If two are non-adjacent, say $v_1 = u_i$ and $v_2 = u_j$, with $i \neq j \pm 1$, then we obtain a $K_{3,3}$ on the vertex set $\{u, v_1, v_2\} \cup \{v, u_{i+1}, u_{j+1}\}$ (see Fig. 7.13(c)). If they are all adjacent, then either there are only two of them, say $v_1 = u_i$ and $v_2 = u_{i+1}$, and it is easy to draw v and all the required edges inside the triangle uu_iu_{i+1}, contradicting the fact that G is not planar; or there are three of them, and the cycle is a triangle $u_1u_2u_3$, so the vertices u, v, u_1, u_2, u_3 form a K_5 (see Fig. 7.14). □

We can now deduce Kuratowski's theorem from Wagner's theorem, although this is not as easy as the other way round. All we have to show is that if H contracts onto K_5 or $K_{3,3}$, then H contains a subdivision of K_5 or $K_{3,3}$. Note,

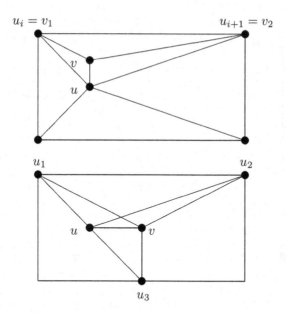

Fig. 7.14 Two more cases in the proof of Wagner's theorem.

however, that the contraction and the subdivision need not be the same. For example, the Petersen graph contracts onto K_5 but does not contain a subdivision of K_5 (Why not?).

Theorem 7.17 *If H is a graph which contracts onto K_5 or $K_{3,3}$, then H contains a subdivision of K_5 or $K_{3,3}$.*

Proof First suppose that H contracts onto $K_{3,3}$. Thus, H contains six connected induced subgraphs H_1, H_2, H_3, L_1, L_2, L_3, say, and edges between each H_i and L_j, for all i and j, but no edges between any H_i and H_j, or between L_i and L_j, for $i \neq j$ (see Fig. 7.15(a)). If we choose a suitable set of nine edges between these subgraphs, then in each subgraph, we want to join up the ends of the appropriate three edges. Whatever way we do this, we end up with three paths meeting at a vertex. The subgraph we have drawn connecting these six vertices is then a subdivision of $K_{3,3}$.

Now suppose that H contracts onto K_5, and that the subgraphs corresponding to the vertices of the K_5 are $H_1, \ldots, H_5$. In each H_i there are now two possibilities as to how the four incoming edges can be joined up. Either we obtain four paths all meeting at a single point, or we have two paths meeting at a vertex v, and the other two meeting at a vertex w, joined by a path from v to w. If all five subgraphs have the four paths meeting at a single point, then we have a subdivision of K_5. Otherwise, we have a subgraph consisting of v and w from H_1, say, and the whole of H_2, H_3, H_4 and H_5, which contracts onto $K_{3,3}$, so we are back in the first case (see Fig. 7.15(b)). □

(a)

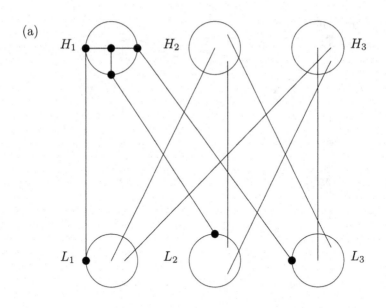

(b)

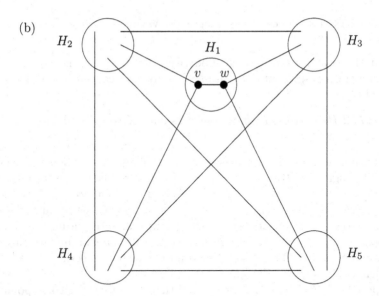

Fig. 7.15 Two cases in the proof of Theorem 7.17. (a) H contracts onto $K_{3,3}$ and (b) H contracts onto K_5.

Exercises

Exercise 7.1 Suppose that two graphs G_1 (with p_1 vertices and q_1 edges) and G_2 (with p_2 vertices and q_2 edges) are homeomorphic (i.e. they are both subdivisions of the same graph H). Prove that $p_1 + q_2 = p_2 + q_1$.

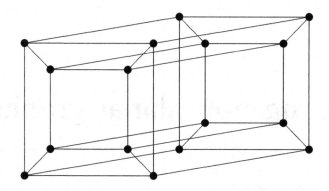

Fig. 7.16 The four-dimensional cube.

Exercise 7.2 The graph of the four-dimensional cube is made by taking two copies of the three-dimensional cube and joining corresponding vertices by edges, as shown in Fig. 7.16. Determine whether this graph is planar. If it is, draw it in the plane. If it is not, try to draw it on a torus.

Exercise 7.3 A planar graph is called **maximal** if no edge can be added to it without violating planarity. Show that any maximal planar graph can be embedded in the plane so that every region has three edges. Find a maximal planar graph which is not a complete graph.

Exercise 7.4 Show that any 5-connected **planar** graph has at least 12 vertices. Give an example of a 5-connected planar graph.

Exercise 7.5 Does there exist a 6-connected planar graph? Justify your answer.

Exercise 7.6 Prove that the complete graphs are the only connected graphs which cannot be disconnected by removing vertices.

Exercise 7.7 Prove that the Petersen graph does not contain a subdivision of K_5.

Exercise 7.8 Determine (with proof) the connectivity of

1. König's graph (Fig. 5.5);
2. the Herschel graph (Fig. 5.7);
3. the Petersen graph;
4. the complete bipartite graph $K_{m,n}$.

8

Colouring non-planar graphs

8.1 Brooks' theorem

Whilst we cannot prove that every map can be 4-coloured, we can prove that (with one exception) every **triangular** map can be 3-coloured. As usual, we state and prove this in the dual form. In fact, this result is true even without the assumption that the graph is planar, and is the case $\Delta = 3$ of Theorem 8.1.

In a sense, we can regard this and similar results as first steps in the generalization of the four-colour theorem to non-planar graphs. First we note that if a graph G has maximum degree $\Delta = \Delta(G)$, then its vertices can be coloured with at most $\Delta + 1$ colours. To prove this, we simply colour the vertices one at a time. At each stage, the vertex is joined to at most Δ of the vertices which have already been coloured, so can be coloured itself.

The following theorem, due to Brooks, says that with two exceptions, we can manage with one colour fewer. The exceptions are genuine: $K_{\Delta+1}$ has maximum degree Δ, but requires $\Delta + 1$ colours. Similarly, if $\Delta = 2$, and G is a cycle of odd length, then we are forced to colour with alternate colours round the cycle, which produces a 2-colouring if and only if the cycle has even length.

Theorem 8.1 (Brooks) *Any connected graph G with maximum degree Δ can be Δ-vertex-coloured, unless G is isomorphic to $K_{\Delta+1}$, or to a cycle of odd length in the case $\Delta = 2$.*

Proof Since a cycle of even length can be 2-coloured, we assume from now on that $\Delta \geqslant 3$. We may also assume that G is 2-connected, since otherwise there is a cutpoint, and we can colour the two blocks separately, and then match up the colouring on the point which is in both blocks.

Moreover, we may assume that **every** vertex v of G has degree Δ, for if not, choose a vertex of degree strictly less than Δ, and remove it from the graph: by induction on the number of vertices, we can colour this smaller graph with at most Δ colours, and then colour v with a colour not used by its at most $\Delta - 1$ neighbours. We divide into two cases according to whether G is 3-connected or not.

If G is 3-connected, with n vertices, say, and G is not K_n, then we can choose a vertex v_n with two non-adjacent neighbours v_1 and v_2 (Fig. 8.1). We now choose recursively v_{n-1}, v_{n-2} and so on, such that each v_i is adjacent to at least one of $v_{i+1}, \ldots, v_n$. This can be done since otherwise v_1 and v_2 would

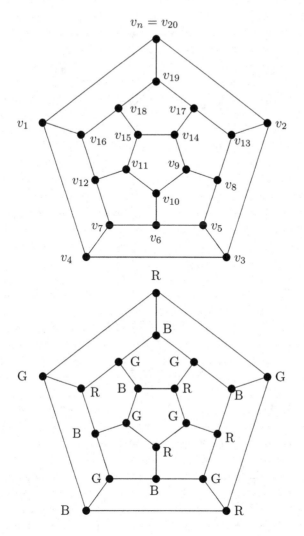

Fig. 8.1 A 3-connected case of Brooks' theorem, $\Delta = 3$.

disconnect $\{v_{i+1}, \ldots, v_n\}$ from the rest of the graph, contradicting the fact that G is 3-connected. Now we can colour v_1, v_2 with the same colour, and then colour $v_3, v_4, \ldots$ in turn, since each v_i adjacent to at most $\Delta - 1$ of the previous vertices $v_1, \ldots, v_{i-1}$. Finally, v_n is adjacent to v_1 and v_2, which have the same colour, and to at most $\Delta - 2$ other vertices, so v_n can also be coloured.

If G is 2-connected but not 3-connected, then we order the vertices in a slightly different way (Fig. 8.2). We first let $\{u, v_n\}$ be a pair of vertices which disconnects the graph. Now the graph obtained from G by removing v_n and all edges incident to v_n is 1-connected but not 2-connected, so has at least two blocks. Moreover, at least two of these blocks are 'endblocks'—which means that

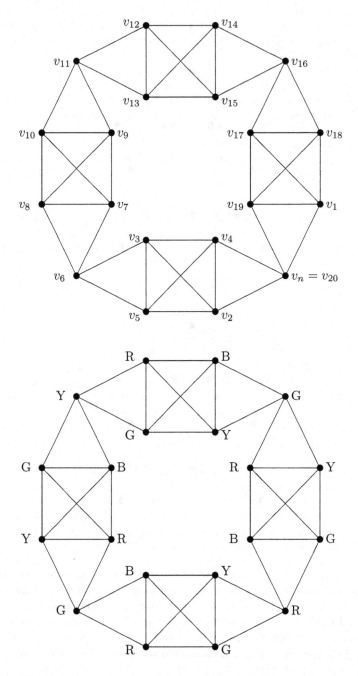

Fig. 8.2 A non-3-connected case of Brooks' theorem, $\Delta = 4$.

they contain only one cutvertex of the graph. Now v_n is adjacent to at least one vertex (other than the cutvertex) in every endblock, for otherwise, the cutvertex is a cutvertex for the whole of G, contradicting the fact that G is 2-connected. If we choose such vertices v_1 and v_2 in two different endblocks, then clearly v_1 and v_2 are non-adjacent.

Now we choose $v_{n-1}, v_{n-2}, \ldots$ as in the 3-connected case, and apply the same colouring algorithm, with the same result. (Note that we need $\Delta \geqslant 3$ to show that v_n has degree at least 3, and hence is joined to a vertex other than v_1 and v_2.) □

The above proof of Brooks' theorem is due to Lovasz. A different proof of Brooks' theorem, using Kempe chains, is given by Robin Wilson [53]. The original proof of Brooks [15] is different again.

8.2 The chromatic number

Having spent a lot of effort trying to decide whether planar graphs can always be coloured with four colours, it is natural to ask in general how many colours a given graph needs. We define the **chromatic number** of a graph G to be the minimum number of colours required to colour the vertices of G, such that adjacent vertices are coloured differently. We write $\chi(G)$ for the chromatic number, and if $\chi(G) = n$ we say G is n-**chromatic** (as opposed to n-colourable, which means $\chi(G) \leqslant n$).

For example, the 1-chromatic graphs are exactly the graphs with no edges, and the 2-chromatic graphs are exactly the bipartite graphs.

In general, there are not very good bounds on the chromatic numbers of graphs. We have proved Brooks' theorem, which says that with the exception of complete graphs and odd cycles, $\chi(G) \leqslant \Delta(G)$, where $\Delta(G)$ denotes the maximum degree of the vertices of G. The following result is similar in flavour. Recall that $\delta(G)$ denotes the minimum degree of the vertices of G.

Theorem 8.2 $\chi(G) \leqslant 1 + \max\{\delta(H)\}$, where the maximum is taken over all induced subgraphs H of G.

Proof The result is clear if $\chi(G) = 1$, that is, G has no edges. Now suppose that $\chi(G) = n \geqslant 2$, and choose H an induced subgraph with $\chi(H) = n$, and with H as small as possible (i.e. H has as few vertices as possible). Note that we may have $H = G$.

Removing any vertex v from H leaves a graph colourable with $n - 1$ colours, by our choice of H, so v must have degree at least $n - 1$ in H, in order to force us to use the nth colour. But this is true for all vertices v in H, so $\delta(H) \geqslant n - 1$. Finally, $\delta(H)$ for this particular H is less than or equal to the maximum of $\delta(H)$ over **all** induced subgraphs, and the result follows. □

Two vertices are called **independent** if they are not adjacent. We define the **vertex independence number** of a graph to be the maximum number of vertices in the graph with the property that no two are adjacent, in other words, the maximum number of mutually independent vertices. Write $\alpha(G)$ for the vertex independence number of G.

Note that the set of vertices coloured in any one colour must have this property, so there must be at least $p/\alpha(G)$ different colours. On the other hand, if you choose $\alpha(G)$ mutually non-adjacent vertices, and colour them in one colour, then you need at most $p - \alpha(G)$ colours for the remaining $p - \alpha(G)$ vertices, giving at most $p - \alpha(G) + 1$ colours in all. Thus, we have proved the following.

Theorem 8.3 $p/\alpha(G) \leqslant \chi(G) \leqslant p - \alpha(G) + 1$.

In general, these bounds are not very good, although they are 'best possible' in the sense that there are graphs which meet each of them. For example, complete graphs meet both. To see this, observe that for a complete graph K_p on p vertices, $\alpha(K_p) = 1$, since any pair of vertices is adjacent. Thus,

$$\frac{p}{\alpha(K_p)} = p = p - \alpha(K_p) + 1.$$

Therefore, $\chi(K_p) = p$, which of course we knew already.

The following bounds consider the graph and its complement simultaneously.

Definition 8.4 *The **complement** $\overline{G}$ of a graph G has the same vertex set, but the complementary set of edges. In other words, vertices u and v are adjacent in $\overline{G}$ if and only if they are not adjacent in G.*

For example, the complement of the complete graph K_n is the graph on n vertices with no edges. You might expect that if $\chi(G)$ is small, then it cannot have too many edges, so that $\overline{G}$ would have lots of edges, and therefore, $\chi(\overline{G})$ is large. Broadly speaking this is true, and the following result, due to Nordhaus and Gaddum, makes this more precise.

Theorem 8.5 $2\sqrt{p} \leqslant \chi(G) + \chi(\overline{G}) \leqslant p + 1$, and

$$p \leqslant \chi(G) \cdot \chi(\overline{G}) \leqslant \left(\frac{p+1}{2}\right)^2.$$

Proof We shall prove that $p \leqslant \chi(G) \cdot \chi(\overline{G})$ and $\chi(G) + \chi(\overline{G}) \leqslant p + 1$. The other two inequalities will then follow immediately from the fact that the arithmetic mean of two positive numbers is greater than or equal to the geometric mean, that is

$$\frac{a+b}{2} \geqslant \sqrt{ab}$$

for positive real numbers a and b.

The first result follows from the result we have already proved in Theorem 8.3, that $p/\alpha(G) \leqslant \chi(G)$. For G contains a set of $\alpha(G)$ mutually non-adjacent vertices, so that $\overline{G}$ contains the complete graph on these same vertices, and therefore, $\chi(\overline{G}) \geqslant \alpha(G)$. Hence

$$\chi(G) \cdot \chi(\overline{G}) \geqslant \chi(G)\alpha(G) \geqslant p$$

as required.

The second result is proved by induction on the number of vertices, p. Pick any vertex v in G, and let H be the graph obtained by removing it and its incident edges. Then $\overline{H}$ is the graph obtained by removing v from $\overline{G}$. By the induction hypothesis $\chi(H) + \chi(\overline{H}) \leqslant p$, since H has $p - 1$ vertices. Let d denote the degree of v in G, so that the degree of v in $\overline{G}$ is $p - 1 - d$.

If $d < \chi(H)$, then colour H with at most $\chi(H)$ colours, and then v has fewer than $\chi(H)$ neighbours in G, so we can colour v in G with one of these same $\chi(H)$ colours. Therefore, $\chi(G) = \chi(H)$, and clearly $\chi(\overline{G}) \leqslant \chi(\overline{H}) + 1$, so

$$\chi(G) + \chi(\overline{G}) \leqslant \chi(H) + \chi(\overline{H}) + 1 \leqslant p + 1$$

as required.

Similarly, if $p - 1 - d < \chi(\overline{H})$ we use the same argument in the complementary graphs. The only remaining case is when $d \geqslant \chi(H)$ and $p - 1 - d \geqslant \chi(\overline{H})$. Adding together these two inequalities we obtain $p - 1 \geqslant \chi(H) + \chi(\overline{H})$, so

$$p \geqslant \chi(H) + \chi(\overline{H}) + 1 \geqslant \chi(G) - 1 + \chi(\overline{G}),$$

and therefore $\chi(G) + \chi(\overline{G}) \leqslant p + 1$ again. $\qquad\square$

Remark As we noted in the proof, it is obvious that $\chi(\overline{G}) \geqslant \alpha(G)$, since there is a set of $\alpha(G)$ mutually non-adjacent vertices in G, which means that the corresponding vertices in $\overline{G}$ induce a complete graph, so must be coloured different colours. Thus, the second inequality $\chi(G) + \chi(\overline{G}) \leqslant p + 1$ in Theorem 8.5 implies the second inequality $\chi(G) \leqslant p - \alpha(G) + 1$ in Theorem 8.3.

A more precise expression of the feeling that the more edges there are, the more colours you need, is given by the inequality $\chi(G) \geqslant p^2/(p^2 - 2q)$. At one extreme, when there are no edges, so $q = 0$, this gives $\chi(G) \geqslant 1$, which is at least true, if not very helpful. At the other extreme, if G is a complete graph then $q = p(p-1)/2$, so the inequality reduces to $\chi(G) \geqslant p$. Thus, the inequality generalizes some results we already know.

Theorem 8.6

$$\chi(G) \geqslant \frac{p^2}{p^2 - 2q}.$$

Proof Suppose that G is coloured with k colours, and the numbers of vertices of each colour are $p_1, \ldots, p_k$, say. Let n be the number of edges in the complement $\overline{G}$. Then, clearly, all vertices of one colour are joined in $\overline{G}$, so

$$n \geqslant \binom{p_1}{2} + \binom{p_2}{2} + \cdots + \binom{p_k}{2}$$

$$= \frac{1}{2} \sum_{i=1}^{k} p_i^2 - \frac{1}{2} \sum_{i=1}^{k} p_i$$

$$\geqslant \frac{1}{2k} \left(\sum_{i=1}^{k} p_i \right)^2 - \frac{1}{2} \sum_{i=1}^{k} p_i$$

by the Cauchy–Schwartz inequality applied to the two vectors $(p_1, \ldots, p_k)$ and $(1, \ldots, 1)$. More precisely, these two vectors have norms $\sum_{i=1}^{k} p_i^2$ and k respectively, and inner product $\sum_{i=1}^{k} p_i$, and the square of the inner product is at most the product of the norms, that is $(\sum_{i=1}^{k} p_i)^2 \leqslant k \sum_{i=1}^{k} p_i^2$.

Also, we have that $n + q = p(p-1)/2$ and $\sum_{i=1}^{k} p_i = p$, so

$$\frac{p(p-1)}{2} = n + q \geqslant \frac{1}{2k} p^2 - \frac{1}{2} p + q$$

$$\Rightarrow p^2 \geqslant \frac{p^2}{k} + 2q$$

$$\Rightarrow p^2 - 2q \geqslant \frac{p^2}{k}$$

$$\Rightarrow k \geqslant \frac{p^2}{p^2 - 2q}$$

(since $q \leqslant p(p-1)/2$ implies that $p^2 - 2q$ is positive), as required. $\qquad\square$

An inequality in the other direction is the following.

Theorem 8.7

$$\chi(G) \leqslant 1 + \sqrt{\frac{2q(p-1)}{p}}.$$

Proof Suppose that $\chi(G) = k$ and choose a k-colouring with the maximum possible number, say n_1, of vertices of the first colour χ_1. Let S_1 be such a set of n_1 vertices. Then every vertex x not in S_1 must be joined to a vertex in S_1, for otherwise, x could be coloured with the colour χ_1, contradicting the maximality of n_1.

Similarly, among these colourings choose one with the maximum possible number, say n_2, of vertices of the second colour χ_2, and choose such a set, S_2 say, of n_2 vertices. Then every vertex not in S_1 or S_2 is joined to a vertex in S_2, by the maximality of n_2.

Continuing in this way, every vertex not in $S_1 \cup \cdots \cup S_i$ must be joined to at least one vertex in each of $S_1, \ldots, S_i$. Thus we construct at least $n_2 + 2n_3 + \cdots + (k-1)n_k$ distinct edges, so

$$q \geqslant n_2 + 2n_3 + \cdots + (k-1)n_k$$

$$\geqslant 1 + 2 + \cdots + (k-1)$$

$$= \frac{k(k-1)}{2}$$

$$= \frac{(k-1)^2}{2} \cdot \frac{k}{k-1}$$

$$\geqslant \frac{(k-1)^2}{2} \cdot \frac{p}{p-1}$$

since $k \leqslant p$. Therefore $(k-1)^2 \leqslant 2q(p-1)/p$, and substituting $\chi(G)$ for k we obtain

$$\chi(G) - 1 \leqslant \sqrt{\frac{2q(p-1)}{p}}$$

as required. $\qquad\qquad\qquad\qquad\qquad\qquad\qquad\qquad\qquad\qquad\qquad\square$

8.3 Hadwiger's conjecture

We next discuss a conjecture which can be thought of as a generalization of the four-colour conjecture. First, we restate the four-colour conjecture in the contrapositive: every 5-chromatic graph G is non-planar. Then Wagner's theorem (Theorem 7.2) implies that G has a subgraph which contracts onto K_5 or $K_{3,3}$. But $K_{3,3}$ is 2-colourable, so you might expect this case not to arise. It has indeed been proved (using the four-colour theorem) that every 5-chromatic graph has a subgraph which contracts onto K_5.

Hadwiger's conjecture (enunciated in 1943) is that this holds more generally: every n-chromatic graph has a subgraph which contracts onto K_n. This has not been proved in general. Indeed, the cases $n \geqslant 7$ are all still open. The cases $n \leqslant 3$ are trivial, the case $n = 4$ was proved by Hadwiger himself [25] (see also Theorem 8.8), and the case $n = 5$ was proved by Wagner to be equivalent to the four-colour theorem. The case $n = 6$ was proved only recently, by Robertson, Seymour and Thomas [41], again using the four-colour theorem.

We conclude this section with a proof of the conjecture for $n = 4$. That is, we prove that if G is any 4-chromatic graph, then G has a subgraph which contracts onto K_4. Restating this in the contrapositive, we have the following.

Theorem 8.8 *If G is a graph which has no subgraph contracting onto K_4, then G is 3-colourable.*

Proof We prove this by induction on the number of vertices of G. We may assume that G is connected, and since trees are 2-colourable, we may assume that G contains a cycle. We choose a cycle $C = v_1 v_2 \cdots v_k v_1$ of minimal length (so that k is the girth of G), and let H be the induced subgraph on all the remaining vertices other than $v_1, v_2, \ldots, v_k$. Let $H_1, H_2, \ldots, H_r$ be the components of H.

We claim that, for each i, at most two of the vertices in the cycle C are adjacent to vertices of H_i. For if three vertices in C are adjacent to H_i, then we can contract C to a triangle, and contract H_i to a vertex, such that the corresponding subgraph of G contracts onto K_4 (see Fig. 8.3). This contradiction proves the claim.

Now if H_i is adjacent to just one vertex in C, this is a cutvertex of G, and a 3-colouring of G can be obtained from 3-colourings of the two pieces into which G is cut. Thus, we may assume that each H_i is adjacent to exactly two vertices in C. For the rest of this proof we will call these vertices the 'feet' of H_i.

Next we claim that the feet of H_i cannot straddle the feet of H_j. For otherwise, the subgraph on C, H_i and H_j can be contracted onto the four feet, giving a K_4 as in Fig. 8.4.

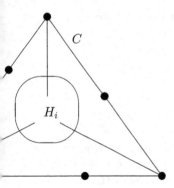

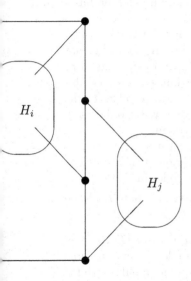

ith three feet in the proof of Theorem 8.8.

ith stradding feet in the proof of Theorem 8.8.

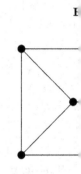

Fig. 5.2

and the cycle C, we encounter a series of nested in Fig. 8.5.

t of H_i, then the outermost H_i can be redrawn g two nests. Therefore, we may assume there here is only one H_i, in which case $H = H_i$ is

ide the graph into two pieces G_1 and G_2, whose v and w on C. Adding the edge vw to each of eady, we can 3-colour the two resulting graphs, a 3-colouring of G by matching the colours of v

In his 1880 pa
have Hamilton cyc
cannot have a brid
this prevents the
bridge—once this
to where we starte
was tacitly assumi
he gives a bridgele
a multigraph, not
that the graph be
as being unimport

Other counter
given by Petersen
and Fig. 5.5). Th
provided a counte
vertices and 25 cc

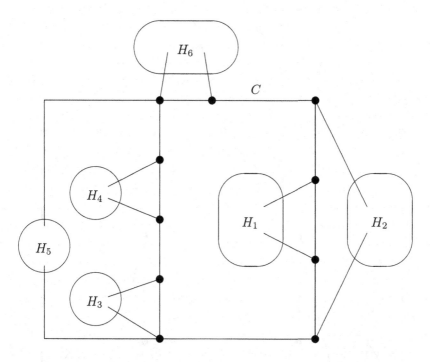

Fig. 8.5 Three nested collections of H_i in the proof of Theorem 8.8.

In the second case, there is a vertex x in C which is not adjacent to any vertex in H. Let G_1 be the graph obtained by removing x and the two incident edges from G. By induction, G_1 is 3-colourable. But x is adjacent to just two vertices in G, so can be coloured with the third colour. $\qquad\square$

8.4 The Hajós conjecture

A variant of the Hadwiger conjecture, due to Hajós (pronounced 'hoy-oash'), has subdivision instead of contraction. That is to say, the Hajós conjecture is that if a graph G is n-chromatic, then G has a subgraph which is a subdivision of K_n. Although this conjecture sounds superficially similar to the Hadwiger conjecture, it turns out to be quite different. For $n \leqslant 3$ the conjecture is easily seen to be true, while for $n = 4$ it was proved by Dirac [20] in 1952. The general case remained unsolved for 35 years. However, in general it is false, and counterexamples are now known for all $n \geqslant 7$. Thus, for all $n \geqslant 7$ there exists an n-chromatic graph which contains no subdivision of K_n. The cases $n = 5$ and 6 are still unresolved. It is clear that, if true, the case $n = 5$ must be hard, as it would imply the four-colour theorem.

Next we construct Catlin's counterexample to Hajós's conjecture (see [16]). It may be obtained by taking five complete graphs K_3, K_2, K_3, K_3 and K_2 in cyclic order, and joining every vertex of one such graph to every vertex of the next one in the cyclic order. A picture is shown in Fig. 8.6.

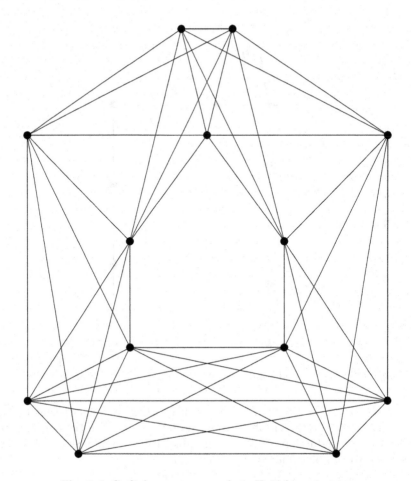

Fig. 8.6 Catlin's counterexample to Hajós's conjecture.

Notice that the bottom two copies of K_3 in the figure are joined together to form a K_6. Notice also that any vertex outside this K_6 has at most five internally disjoint paths from it into the K_6. In particular, there is no subgraph which is a subdivision of K_7. On the other hand, there is no 6-colouring of the graph: for if there were, then the two adjacent copies of K_3 would (without loss of generality) require colours 1, 2, 3 and 4, 5, 6 respectively, and then the copies of K_2 on either side would need (without loss of generality) colours 4, 5 and 1, 2 respectively. But then the last K_3 would require three colours distinct from $1, 2, 4, 5$, which is impossible. Thus, we have shown that Catlin's graph is a counterexample to Hajós's conjecture for $n = 7$.

Now if we add a vertex, joined to all of the original vertices, then we increase the chromatic number by 1. Moreover, if the new graph contains a subdivision of K_n, then the old one contains a subdivision of K_{n-1}. Therefore, we obtain counterexamples for $n = 8, 9, \ldots$ by adding one vertex at a time in this way.

8.5   The chromatic polynomial

We turn now to a different aspect of colouring. Rather than asking for the minimum number of colours we can get away with, suppose we are given a set of t colours, how many different ways are there of colouring a given graph with these t colours? Clearly, this is a function both of the graph G and of the number t, and we write this function as $P(G, t)$. It was introduced by Birkhoff [11] in 1912, in the dual form, for maps, in the hope that it would help solve the four-colour problem: for a graph G is 4-colourable if and only if $P(G, 4) > 0$, in other words if and only if 4 is not a root of $P(G, t)$.

For certain graphs this function is easy to calculate. For example, if the graph has n vertices and no edges, then each vertex can be coloured independently of the rest, with one of the t colours, and therefore, there are t^n possible colourings altogether. Again, for the complete graph K_n, we can colour the first vertex with any of the t colours, the next with any of the remaining $t - 1$, and so on, giving $P(K_n, t) = t(t-1)(t-2) \cdots (t-n+1)$. Similarly, if G is a tree on p vertices, then it is easy to see that $P(G, t) = t(t - 1)^{p-1}$. This is because the first vertex can be coloured in any of t colours, and each adjacent vertex in any of the remaining $t - 1$ colours, and so on until all vertices are coloured.

Note also that if G is the disjoint union of two graphs H and K, then the number of colourings of G is the product of the number of colourings of H with the number of colourings of K. That is, $P(G, t) = P(H, t) \cdot P(K, t)$. Thus we may restrict attention to connected graphs.

If G is any connected graph which is not a complete graph, we can choose two non-adjacent vertices u and v. These can either be coloured the same colour, or different colours. If they are coloured differently, we might as well add the edge uv, as this does not change the number of such colourings. If they are coloured the same, we might as well replace them by a single vertex, joined to all the vertices which were joined to either u or v (or both). Thus $P(G, t) = P(H_1, t) + P(H_2, t)$, where H_1 has more edges than G (but the same number of vertices), and H_2 has fewer vertices than G. (Notice incidentally that H_2 is the contraction of H_1 along the edge uv.)

Now any combination of adding edges and identifying pairs of vertices in a connected graph must eventually lead us to a complete graph. This means that the process above leads to an expression for $P(G, t)$ as a sum of $P(K_n, t)$ for various (not necessarily distinct) values of n. In particular, we have proved that $P(G, t)$ is always a polynomial in t. It is called the **chromatic polynomial**.

For example, the graph G in Fig. 8.7 is K_4 with an edge removed. So H_1 is K_4, obtained by putting the edge uv back, while H_2 is K_3, obtained by replacing u and v by a single vertex. Therefore,

$$
\begin{aligned}
P(G, t) &= P(K_4, t) + P(K_3, t) \\
&= t(t - 1)(t - 2)(t - 3) + t(t - 1)(t - 2) \\
&= t(t - 1)(t - 2)^2.
\end{aligned}
$$

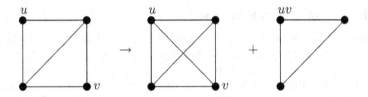

Fig. 8.7 Calculating a chromatic polynomial.

In some cases it may be easier to use the formula the other way round:

$$P(G,t) = P(G_1,t) - P(G_2,t),$$

where G_1 is obtained from G by removing an edge e, and G_2 is obtained by contracting the same edge to a single vertex. In this form, the formula is known as the **deletion–contraction formula**. For example, if G is a cycle, we can remove one edge at a time to prove the following.

Proposition 8.9 *If G is a cycle with p vertices, then* $P(G,t) = (t-1)^p + (-1)^p(t-1)$.

Proof By induction on the number of vertices. If $p = 3$, then G is the complete graph K_3, so $P(G,t) = t(t-1)(t-2) = (t-1)^3 - (t-1)$, as required. For $p > 3$, removing an edge gives a tree, which has chromatic polynomial $t(t-1)^{p-1}$, and contracting it gives a cycle on $p-1$ vertices, which by induction has chromatic polynomial $(t-1)^{p-1} - (-1)^p(t-1)$. Therefore,

$$\begin{aligned}P(G,t) &= t(t-1)^{p-1} - (t-1)^{p-1} + (-1)^p(t-1) \\ &= (t-1)^p + (-1)^p(t-1)\end{aligned}$$

as required.   □

Proposition 8.10 *If G is the union of two subgraphs H and K, such that the intersection of H and K is a complete graph on k vertices, then*

$$P(G,t) = \frac{P(H,t) \cdot P(K,t)}{t(t-1)\cdots(t-k+1)}.$$

Proof First, we can colour H in any of $P(H,t)$ ways. Then we have chosen one of the $t(t-1)\cdots(t-k+1)$ possible colourings of $H \cap K$. Since all colourings of this complete graph extend to the same number of colourings of K, we have $P(K,t)/t(t-1)\cdots(t-k+1)$ ways of extending this particular colouring to the whole of G.   □

Another result which is sometimes useful for calculations is the following.

Proposition 8.11 *If G is obtained from a subgraph H by adjoining one vertex v joined to all vertices of H, then $P(G,t) = t \cdot P(H,t-1)$.*

Proof Colour v in any of t ways, and then H has to be coloured using the remaining $t-1$ colours.   □

For example, Proposition 8.9 tells us that a cycle on $p-1$ vertices has chromatic polynomial $(t-1)^{p-1}+(-1)^{p-1}(t-1)$ so a wheel on p vertices has chromatic polynomial $t(t-2)((t-2)^{p-2}+(-1)^{p-1})$.

Some properties of the chromatic polynomial are fairly straightforward to prove, and enable you to read off some information about the graph from its chromatic polynomial.

Proposition 8.12 $P(G,t)$ is a monic polynomial of degree p (the number of vertices in G).

Proposition 8.13 The coefficient of t^{p-1} in $P(G,t)$ is $-q$ (where q is the number of edges in G).

Proof Using the above notation we have $P(G,t) = P(H_1,t) + P(H_2,t)$, where H_1 has the same number of vertices as G, while H_2 has one fewer. Therefore, the coefficient of t^{p-1} in $P(H_2,t)$ is 1. So every time we remove an edge from H_1, we add 1 to the coefficient of t^{p-1}. This continues until we have no edges left, when the chromatic polynomial is t^p. □

Proposition 8.14 The constant term of $P(G,t)$ is 0.

Proof Put $t = 0$, and get the number of colourings with no colours, which is 0.

Alternatively, note that you can always choose the first colour arbitrarily, so there is a factor of t in the polynomial. □

Proposition 8.15 The trailing term in $P(G,t)$ is the term in t^k, where k is the number of connected components of G.

Proof It is sufficient to prove the case $k = 1$, since $P(G,t)$ is the product of $P(C_i,t)$ over the components C_i of G. One half of the proof is obvious: the previous proposition shows that the constant term is 0. So we only need to prove that the t-term is non-zero. This is a consequence of the following more precise result. □

Lemma 8.16 Let G be a connected graph with n vertices. Then the coefficient of t in $P(G,t)$ is positive if n is odd, and negative if n is even.

Proof We prove this by induction on n. For $n = 1$, we have $P(G,t) = t$, and the result holds. Suppose now that $n > 1$, and suppose the result holds for all graphs with $n-1$ vertices. Pick a vertex, v, of degree d, say, in G, and let $w_1, \ldots, w_d$ be the neighbours of v. Now let G_i be the graph obtained by removing the edges $vw_1, \ldots, vw_i$. Thus $G_0 = G$, and G_d is a graph consisting of the single vertex v, together with a graph H on $n-1$ vertices.

Now our basic lemma for calculating chromatic polynomials (the deletion–contraction formula) gives $P(G_i,t) = P(G_{i-1},t) + P(H_i,t)$, where H_i is some graph on $n-1$ vertices. By induction on i we obtain

$$P(G_d,t) = P(G_0,t) + \sum_{i=1}^{d} P(H_i,t) = P(G,t) + \sum_{i=1}^{d} P(H_i,t).$$

But $P(G_d, t) = tP(H, t)$, so has zero t-term, while our inductive hypothesis implies that all the t-terms of the $P(H_i, t)$ have the same sign. Therefore, the t-term of $P(G, t)$ has the opposite sign, and the induction continues. □

More generally, we have:

Theorem 8.17 *If G has p vertices, q edges and k components, then*

$$P(G, t) = t^p - qt^{p-1} + \cdots + (-1)^{p-j} a_j t^j + \cdots + (-1)^{p-k} a_k t^k,$$

where each $a_j > 0$, for $k \leqslant j \leqslant p$.

Proof By induction on the number of edges, using the deletion–contraction formula for the chromatic polynomial, $P(G, t) = P(G_1, t) - P(G_2, t)$, where G_1 is obtained by removing an edge, and G_2 is obtained by contracting the same edge, from G. The induction starts with no edges, when $P(G, t) = t^p$ and $p = k$, so the result holds trivially. Now if G has at least one edge, then both G_1 and G_2 have fewer edges, so by induction we can assume the result for them. Moreover, G_1 has p vertices, $q - 1$ edges, and either k or $k + 1$ components, while G_2 has $p - 1$ vertices and k components, so we have

$$P(G_1, t) = t^p - (q - 1)t^{p-1} + \cdots + (-1)^{p-j} b_j t^j + \cdots + (-1)^{p-k} b_k t^k$$
$$P(G_2, t) = t^{p-1} + \cdots - (-1)^{p-j} c_j t^j + \cdots - (-1)^{p-k} c_k t^k$$

where all $b_j > 0$ and all $c_j > 0$, except possibly c_k, which might be zero. The result follows by substituting into $P(G, t) = P(G_1, t) - P(G_2, t)$. □

We can actually deduce a little more from the proof of this theorem. With the above notation, we had $a_i = b_i + c_i$, with $c_i \geqslant 0$, so that $a_i \geqslant b_i$ for all i. Here a_i denotes the absolute value of the coefficient of t^i in the chromatic polynomial of G, while b_i denotes the corresponding value in the chromatic polynomial of G_1, obtained from G by removing an edge. If we now remove enough edges to leave just a spanning tree T, say, then we have $a_i \geqslant d_i$, where d_i is the absolute value of the coefficient of t^i in the chromatic polynomial of T. But the chromatic polynomial of T is $t(t - 1)^{p-1}$, so $d_i = \binom{p-1}{i-1}$. Therefore, we have proved the following result.

Theorem 8.18 *If G is a connected graph with chromatic polynomial*

$$P(G, t) = t^p - qt^{p-1} + \cdots + (-1)^{p-j} a_j t^j + \cdots + (-1)^{p-k} a_k t^k,$$

then $a_i \geqslant \binom{p-1}{i-1}$.

In some cases you can actually tell a lot more about the graph from its chromatic polynomial. For example, it is easy to see that a tree on p vertices has chromatic polynomial $t(t - 1)^{p-1}$. Conversely,

Theorem 8.19 *If $P(G, t) = t(t - 1)^{p-1}$ then G is a tree on p vertices.*

Proof $P(G, t)$ has a non-zero t term, so G is connected. Also, the coefficient of t^{p-1} is $-(p-1)$, so G has $p-1$ edges. Therefore G is a tree, by virtue of Lemma 2.8. □

Note that a graph is in general not determined by its chromatic polynomial, as all trees on p vertices have the same chromatic polynomial. Another example is given in Fig. 8.8.

The original purpose of introducing the chromatic polynomial was to study the roots. We know that 0 is always a root, of multiplicity equal to the number of components. Also, if there is at least one edge then 1 is a root. Indeed, if G contains a complete subgraph K_n, then $0, 1, 2, \ldots, n-1$ are all roots of $P(G, t)$. Moreover, since the coefficients alternate in sign, all roots are non-negative. For if $\lambda < 0$ then all terms in $P(G, \lambda)$ have the same sign (positive if p is even, negative if p is odd), so $P(G, \lambda)$ cannot be 0.

Theorem 8.20 *For any graph G, the chromatic polynomial has no root strictly between 0 and 1.*

Proof Since the roots of $P(G, t)$ are roots of $P(C, t)$ for one of the connected components C of G, it suffices to prove the result in the case when G is connected. We prove this case by induction on the number of edges. If G has only one edge, then it has two vertices and $P(G, t) = t(t-1)$, so the result holds.

Now choose an edge e. If removing e disconnects G, then G is the union of two connected graphs G_1 and G_2 whose intersection is the edge e and its two endpoints. Thus, the intersection is a copy of K_2, and we apply Proposition 8.10 to obtain

$$P(G, t) = \frac{P(G_1, t) \cdot P(G_2, t)}{t(t-1)}$$

which is non-zero for all t strictly between 0 and 1, by induction. Moreover, this shows that $P(G, t)$ is positive if and only if p is even.

On the other hand, if removing e does not disconnect G, then use the deletion–contraction formula and the stronger induction hypothesis that if $0 < t < 1$ then $P(G, t) > 0$ if p is odd, and $P(G, t) < 0$ if p is even. □

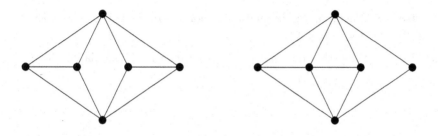

Fig. 8.8 Two graphs with the same chromatic polynomial.

More on the chromatic polynomial can be found in the chapter by Read and Tutte [8, vol. 3].

Exercises

Exercise 8.1 Find some graphs other than complete graphs for which

$$\chi(G) = p - \alpha(G) + 1.$$

Is it possible to find, for each positive integer n, and each positive integer m, such a graph G with $\chi(G) = n$ and $\alpha(G) = m$? Justify your answer.

Exercise 8.2 For each $p \geqslant 1$, and each positive integer $k \leqslant p$, find a graph on p vertices such that $\chi(G) = k$ and $\chi(G) + \chi(\overline{G}) = p + 1$.

Exercise 8.3 For each $p \geqslant 1$, and each positive integer k dividing p, find a graph on p vertices such that $\chi(G) = k$ and $\chi(G)\chi(\overline{G}) = p$.

Exercise 8.4 State the converse of the Hajós conjecture, and show that it is false for all $n \geqslant 3$.

Exercise 8.5 (Hard.) Let G be a minimal counterexample to the Hajós conjecture for $n = 4$, so that G is 4-chromatic but contains no subdivision of K_4.

1. Show that every vertex of G has degree at least 3, and that G has no cutvertex.
2. Show that every vertex of C is connected by a chord (i.e. a path internally disjoint from C) to some other vertex of C.
3. Show that if two such chords intersect then G contains a subdivision of K_4.
4. Show that if no two chords intersect, then G contains a subdivision of K_4.
5. Deduce that the Hajós conjecture is true for $n = 4$.

Exercise 8.6 Calculate the chromatic polynomials of the following graphs:

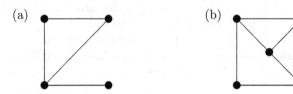

Exercise 8.7 Calculate the chromatic polynomials of the two graphs in Fig. 8.8 and verify that they are equal.

Exercise 8.8 Prove that no graph can have chromatic polynomial $t^4 - 3t^3 + 3t^2$.

Part III

How to prove the four-colour theorem

9
Overview

9.1 Historical remarks

The overall strategy of the proof of the four-colour theorem does not differ greatly from that adopted by Kempe in his 1879 paper. If we consider the vertex-colouring form of the theorem, then the proof goes by induction on the number of vertices. Clearly, the induction starts, as any graph with at most four vertices is 4-colourable.

The general inductive step is to remove a vertex, thereby **reducing** to a smaller case, which by the inductive hypothesis we can assume is 4-colourable. The problem then is to extend the colouring to include the extra vertex, or more generally, to find a way of changing the colouring so that it can be extended to the extra vertex.

As we have seen, if the vertex has degree less than 4, then the colouring extends trivially, while if it has degree 4, we can change the colouring using a Kempe-chain argument in such a way that the new colouring extends to the extra vertex. In modern language, we have shown that a vertex of degree at most 4 is **reducible**. (It would be more logical to say that **any graph containing a vertex of degree at most** 4 is reducible, but the above usage is the one generally adopted.)

The other half of the problem is to show that every planar graph contains one of these reducible configurations. Kempe used Euler's formula to show that every planar graph contains a vertex of degree at most 5. In other words, the graph **cannot avoid** having a vertex of degree 5 or less. We say that this set of 'configurations' (we will define this term more precisely later) is **unavoidable**, since every planar graph contains at least one of them. If we neglect the trivial cases where a vertex has no neighbours or one neighbour, this unavoidable set is as illustrated in Fig. 9.1. Note that a vertex of degree 2 creates a pair of parallel edges, so that strictly speaking we have a multigraph rather than a graph. However, we can easily neglect this case if we like, and restrict our attention to graphs in the strict sense.

The reason why Kempe's attempted proof fails is because the **unavoidable set** here is not a subset of the **reducible set** (i.e. the set of reducible configurations). If we could somehow find an unavoidable set consisting only of reducible configurations, then we would have finished the proof of the four-colour theorem.

Fig. 9.1 Kempe's unavoidable set of configurations.

Historically, the attack was pursued from both sides simultaneously, so that improvements were made in both the unavoidable sets and in the reducibility arguments. It was immediately clear that it was necessary to look not just at individual vertices and their degrees, but at larger configurations of adjacent vertices. And if attempts to prove that a vertex of degree 5 is reducible continued to fail, then it would be necessary to replace this particular configuration in the unavoidable set, by a possibly larger number of larger configurations.

The first result in this direction was Wernicke's proof in 1904 that the vertex of degree 5 could be replaced by a pair of adjacent vertices, one of degree 5 and the other of degree 5 or 6. This then gives extra information about the neighbourhood of the degree 5 vertex, which may be of help when we try to prove reducibility. However, it turned out that this information was quite inadequate for the purpose.

Birkhoff in 1913 approached the problem from the other direction, and showed that a vertex of degree 5 with three consecutive neighbours of degree 5 is reducible. It was clear, though, that there was still a huge gap between the set of known reducible configurations, and any known unavoidable set.

Nevertheless, there was steady progress from both directions. In 1922 Franklin, a student of Birkhoff, improved Wernicke's result by showing that the vertex of degree 5 in the unavoidable set could be replaced by a vertex of degree 5 with two neighbours of degree 5 or 6. At the same time he showed that a vertex of degree 6 with three consecutive neighbours of degree 5 is reducible. Later authors proved reducibility of more and more configurations of this sort.

9.2 Elementary reductions

We know that it is sufficient to consider triangulated graphs—that is, plane graphs in which every face has exactly three edges. This corresponds to our reduction to the case of cubic maps (Theorem 4.6) in the face-colouring version. Triangulated graphs are automatically connected. Moreover, since we already know that vertices of degree at most 4 are reducible, we may assume that our graph contains no vertices of degree less than 5.

There are other reductions of this type, which we shall prove in the next chapter. For example, the neighbours of a vertex of degree d form a cycle of length d, whose interior (or exterior, if the vertex is an exterior vertex) contains a single vertex. Thus, we know that such cycles of length $d \leqslant 4$ are reducible. In Theorem 10.5 and Corollary 10.6 we show by a Kempe-chain argument that **any**

4-cycle is reducible. We can even show (see Theorem 10.7) that any 5-cycle is reducible, **except** for the one troublesome case that we know, namely a 5-cycle consisting of the neighbours of a vertex of degree 5.

Thus, we may if we like assume that G has no separating circuit of length 5 or less, except circuits consisting of the five neighbours of a vertex of degree 5. A graph with this property is called **internally 6-connected**.

9.3 Strategy

From this point on we can assume that our graph is an internally 6-connected triangulation. Various new methods of proving reducibility had been developed over the years, but it always seemed that, whatever unavoidable set was produced, some of its configurations were resistant to all attempts to prove its reducibility. Some method was required to produce unavoidable sets without any such 'bad' configurations.

All methods for producing unavoidable sets rely ultimately on Euler's formula. Specifically, there must be at least 12 vertices of degree 5, and every vertex of degree d bigger than 6 must be compensated for by $(d - 6)$ more vertices of degree 5, by the formula $\sum_v (6 - d(v)) = 12$, which is Proposition 3.18.

Thus, there must be 'plenty' of vertices of degree 5, and the idea is to somehow home in on the places were this plenty is manifested. In other words, **where** (in a suitable sense) do the positive contributions to $\sum_v (6 - d(v))$ come from? The subtlety in this question is, what do we mean by 'where'? If we define 'where' too narrowly, by referring to individual vertices, then the answer is simply that the positive contributions come from the vertices of degree 5, and we have made no progress beyond the point reached by Heawood in 1890. If, on the other hand, we define 'where' too broadly, then we have not located the problems precisely enough, and will never be able to solve them. In the end it turned out that it was sufficient to confine attention to parts of the graph consisting of an n-cycle and its interior, for $n \leqslant 14$. However, this on its own gives no bound on the number of vertices in such a configuration.

The idea of 'discharging' was introduced by Heesch, in order to give a precise meaning to this word 'where'. The idea is to associate a 'charge' of magnitude $6 - d(v)$ to each vertex v, and then to devise a method (called a 'discharging algorithm') of spreading the charge around from one vertex to its neighbours, in such a way that positive charges only arise under certain restricted conditions (where there are 'too many' vertices of degree 5 nearby), which can be identified. These restricted conditions then constitute the 'unavoidable set'.

The way that Appel and Haken were able to prove the four-colour theorem was by successively modifying the discharging algorithm to produce a better unavoidable set each time. By this stage Heesch had developed a feeling for what sort of configurations were likely to be troublesome, and so knew what to look out for when searching for unavoidable sets. Using this intuition, Appel and Haken would look at those configurations in the unavoidable set which looked hard to reduce. Then they would redesign the discharging algorithm to eliminate these particular cases. Of course, this might then introduce new problem cases. However, by repeating this procedure they eventually found a discharging algorithm which produced an unavoidable set of 1936 configurations which they believed they could prove were reducible. These configurations were then proved reducible, with the help of a computer which was programmed by Koch to search for the required colouring extensions.

The proof was completed in 1976. The published proof appears in two parts [4,5]. The latter contains a 63-page table listing all the 1936 configurations. A very readable introduction, describing all the main ideas, can be found in the article by Woodall and Wilson in [8, vol. 1, pp. 83–101].

9.4   Later improvements

It soon transpired that not all of the 1936 configurations were distinct—some were actually repeated, and others were subconfigurations of bigger ones—so that just 1834 of them were actually required in the proof. A little later, further improvements to the proof of unavoidability resulted in another 352 configurations being declared redundant, leaving just 1482 configurations in the unavoidable set.

More recently (in 1997), a simplified version of the Appel–Haken proof has been published by Robertson, Sanders, Seymour and Thomas [43]. It still relies on computer calculations, but the number of unavoidable configurations has been reduced to 633, which are explicitly drawn in the last 9 pages of the article, and the 'discharging algorithm' (used to produce the unavoidable set of configurations— see Chapter 11) is greatly simplified. Indeed, they found an unavoidable set of 591 configurations, but rejected it because it made the proof of reducibility harder. Moreover, the article is much shorter, at 43 pages instead of 139, and gets close to providing a proof which can be checked by the sufficiently determined reader.

10
Reducibility

10.1 The Birkhoff diamond

In the eventual proof of the four-colour theorem, the first step, logically, is to use discharging to produce the unavoidable set, and then the second step is to prove reducibility of each configuration in the unavoidable set. However, historically, many of the reducibility arguments came first, and it is easier to appreciate the subtlety of the discharging arguments if we first see what they are being used for. Indeed, if we know that certain configurations are reducible, we can try to design our discharging algorithm to produce these particular configurations.

As usual, we assume that our graph is triangulated, and that it has no vertices of degree less than 5. For our purposes, a **configuration** inside a plane triangulation G consists of a circuit (called the **boundary ring** of the configuration) together with the part of G in its interior. Note, however, that this is not the definition used by Appel and Haken [4].

As an example to illustrate the nature of these arguments, we consider the 'Birkhoff diamond', that is, a vertex of degree 5 with three consecutive neighbours of degree 5. This is named after G. D. Birkhoff who proved its reducibility in 1913. We draw the induced subgraph on these four vertices and their six neighbours (i.e. the full configuration) in Fig. 10.1. Note that we do not know the degrees of these six neighbours. We also need the following easy lemma. By a **separating triangle**, we simply mean a triangle which separates the graph into an inside and an outside, both of which are non-empty subgraphs.

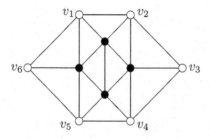

Fig. 10.1 A vertex of degree 5 with three consecutive neighbours of degree 5.

Lemma 10.1 *If G is a minimal counterexample to the four-colour theorem, then G contains no separating triangle.*

Proof   If G contains a separating triangle, then G is the union of two graphs, which intersect just in this triangle. By induction, each of these graphs can be 4-coloured, and in each case the three colours used by the vertices of the separating triangle are different. Thus, by renaming the colours in one of the two subgraphs if necessary, we can match up the two colourings to produce a 4-colouring of the whole graph. This contradiction proves the result.   □

Theorem 10.2 (Birkhoff) *The Birkhoff diamond is reducible.*

Proof   Let G be a graph which contains the Birkhoff diamond, and is as small as possible subject to not being 4-colourable. Now remove the inside of the hexagon, and collapse the hexagon by identifying vertices v_2 and v_4, and joining this vertex to v_6, as in Fig. 10.2. (Note that this is possible, for if v_2 and v_4 were originally adjacent, then there would be a separating cycle of length 3, contrary to Lemma 10.1.) In effect, we contract v_2, v_4 and the four interior vertices to a single vertex $v_{2,4}$.

This gives us a smaller graph, which can be 4-coloured (by our assumption that G is a minimal counterexample). If we colour v_1 red, v_6 green, and $v_{2,4}$ blue, then v_5 can be either yellow or red, while v_3 can be green, yellow or red. Thus, there are now six possibilities for the colouring of the vertices v_1 to v_6, and in five of these cases, we can easily complete this to a 4-colouring inside the hexagon. Indeed, there is a colouring which works in all four of the cases when v_3 is not red (see Fig. 10.3). In the last case, a Kempe-chain argument reduces it to one of the other five cases, as follows.

In this case, if all the red vertices in the boundary ring belong to the same red–yellow chain, we can change the colour of v_4 to green, and complete the new colouring as in Fig. 10.4. If not, and v_3 is in a different red–yellow chain from the other two red vertices, we can recolour it yellow, so it becomes the middle case of the right-hand column. Finally, if v_1 or v_5 is in a different red–yellow chain from the other two, we recolour it to the bottom case in the first column.   □

Remark   The above argument is the one originally given by Birkhoff, in the dual form. See also Barnette [7], and Saaty and Kainen [44]. A similar, perhaps

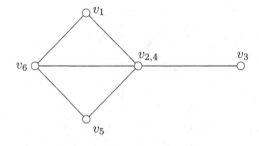

Fig. 10.2 The collapsed hexagon.

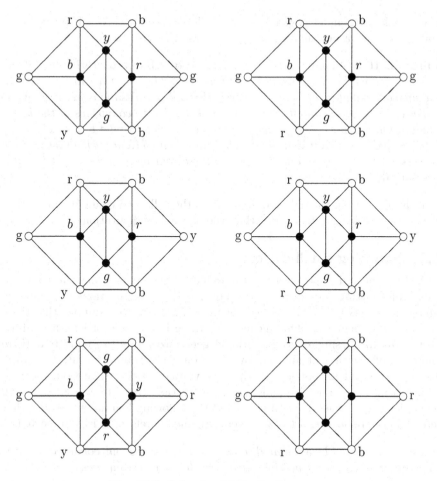

Fig. 10.3 Colouring inside the hexagon.

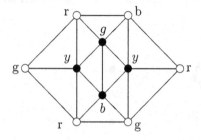

Fig. 10.4 Recolouring inside the hexagon.

slightly easier, argument is given by Woodall and Wilson [8, vol. 1, pp. 83–101], joining v_1 to v_5, rather than v_6 to $v_{2,4}$ (see also Exercise 10.5).

Definition 10.3 *A configuration is called* **B-reducible** *(after Birkhoff) if it can be proved reducible by an argument of this type. That is, if it can be collapsed to a smaller configuration in such a way that all permissible colourings of the smaller configuration can either be directly extended to colourings of the larger configuration, or can be so extended after recolouring a single Kempe chain.*

It is called **A-reducible** *in the special case when no Kempe-chain recolouring is necessary. If more than one Kempe-chain recolouring is necessary, it is called* **C-reducible**.

In this definition, the only requirement on the collapsed configuration is that it has fewer vertices in its interior than the original configuration.

10.2 Reducing small cycles

Birkhoff also proved some other types of reduction theorems, which turned out to be useful in putting general restrictions on possible configurations in a minimal counterexample G to the four-colour theorem. We know, for example, that there is no vertex of degree 4. This means that there is no cycle of length 4 whose interior contains a single vertex. Birkhoff generalized this to saying that there is no cycle of length 4 at all. Moreover, such a cycle would separate the graph into an inside and an outside, so the four vertices of the cycle would form a disconnecting set. Thus, Birkhoff's result says that there is no disconnecting cycle of length 4 (or less). With the aid of the following lemma, we deduce that there is no disconnecting set of four vertices, and therefore, G is 5-connected.

Lemma 10.4 *Let G be a minimal counterexample to the four-colour conjecture. Then, any minimal disconnecting set in G induces at least a cycle.*

Proof If S is a minimal disconnecting set which does not induce a cycle, and S disconnects G into L and R, then we can draw G in the plane in such a way that L and R lie on the left and right of S, respectively (see Fig. 10.5).

Now choose a vertex v in the exterior boundary of L, and a vertex w in the exterior boundary of R. The edge vw can now be added to G without violating

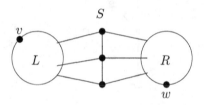

Fig. 10.5 Proof of Lemma 10.4.

planarity. This contradicts the fact that G is a maximal planar graph (i.e. a triangulation). □

Theorem 10.5 *If G is a minimal counterexample to the four-colour conjecture, then G is 5-connected.*

Proof We give a proof due to Saaty and Kainen [44]. We assume G is not 5-connected, and try to obtain a contradiction. By the lemma, there is either a separating triangle or a separating quadrangle. In the former case, as in Lemma 10.1, we can 4-colour the inside and outside separately, and relabel the colours in one part so that the three colours on the vertices of the triangle are the same in both colourings, and we obtain a 4-colouring of the whole graph.

In the latter case, label the vertices of the quadrangle w, x, y, z in order. Add the edge yw to the outside graph—this gives a plane graph, G_1 say, with a smaller number of vertices than G (see Fig. 10.6). Therefore, by induction, G_1 can be 4-coloured, and w, x, y are three different colours. Similarly for the inside of the separating quadrangle—we add the edge wy to the graph consisting of the quadrangle and its interior—this gives a graph G_2 say, which is again a triangulation with fewer vertices than G, so G_2 can be 4-coloured. Then we can match up the colours in the two parts **unless** x and z are coloured the same in one colouring (say the outer one, G_1), and differently in the other (say the inner one, G_2).

If there is no Kempe chain in G_2 from x to z, then we can change the colour of z to match the outer colouring. On the other hand, if there is such a chain, then there is no Kempe chain in G_2 from y to w, so we may change the colour of w in the inner colouring, so that w and y are now coloured the same. Now add the edge xz to the outer graph instead of yw, and colour again. In the new outer colouring, x, y, z are three different colours, and w is either the same colour as y, or a different colour. In the first case, we match the modified inner colouring, while in the second case, we match the original inner colouring. This concludes the proof of the theorem. □

The original proof of Birkhoff is along the same lines, except that instead of triangulating the quadrangle $wxyz$ by introducing the edge yw, he collapses the quadrangle by identifying the two vertices w and y. Thus, the two colourings of the quadrangle $wxyz$ are either $rgrg$ or $rgrb$, say, and the only problem arises when one colouring has $rgrg$ and the other has $rgrb$. Now the usual Kempe-chain argument implies that we can change the first ($rgrg$) colouring to either $rgrb$ or $rgyg$, depending on which Kempe chains exist. In the first case, we match the second colouring, so we may assume the first colouring changes to $rgyg$. Finally, we change the second colouring, by collapsing the quadrangle the other way, identifying x and z. Thus, the second colouring is either $rgrg$ or $rgyg$, say, and matches either the original first colouring, or the new one.

Corollary 10.6 *If G is any planar graph which has a separating cycle of length less than 5, then G is reducible, in the sense that we can prove 4-colourability from 4-colourings of smaller planar graphs.*

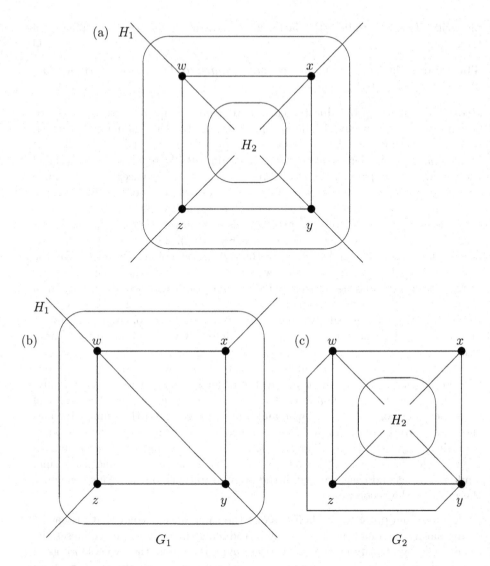

Fig. 10.6 Reducing a separating cycle of length 4. (a) A triangulated graph G consisting of two pieces H_1 and H_2 intersecting in a cycle $wxyzw$ of length 4, (b) G_1 consists of H_1 (including the cycle $wxyzw$), and an extra edge wy, and (c) G_2 consists of H_2 (including the cycle $wxyzw$), and an extra edge wy.

10.3 Birkhoff's reduction theorem

We know that it is impossible to strengthen Theorem 10.5 to saying that if G is a minimal counterexample to the four-colour theorem then G is 6-connected, because G has vertices of degree 5, so can be disconnected by removing the five neighbours of such a vertex. However, it turns out that this is the **only** way of disconnecting G with five vertices. This was also proved by Birkhoff in his 1913 paper [12].

Theorem 10.7 *If G is any planar graph which has a separating circuit C of length 5, such that both the interior and the exterior of C contain at least two vertices of G, then G is reducible.*

Proof We give Birkhoff's original proof in the dual form. As in the proof of Theorem 10.5, we try to colour the inside and outside parts of the graph, and match them up on the cicuit C. The difference now is that we are allowed to add one vertex to each part of the graph, and still have graphs with fewer vertices than G. Let G_1 denote the graph consisting of the cycle C together with its exterior, and let G_2 denote the graph consisting of the cycle together with its interior.

Let G_1' denote the graph obtained from G_1 by adding a vertex joined to all the vertices of C. Then G_1' is a triangulation with fewer vertices than G, so by induction can be 4-coloured. Moreover, this colouring uses exactly three colours on the cycle C, since the fourth colour is required for the central vertex. The colouring of C is therefore of the type (r, g, b, g, b) with one vertex (which we call the **marked vertex**) of one colour, and two each of two other colours.

Now we do the same thing with the interior graph G_2. Let G_2' denote the graph obtained from G_2 by adding a vertex joined to all the vertices of C. We obtain a 4-colouring of G_2' which again induces a 3-colouring on C. The trick now is to try to match up the two colourings of C, so that by putting the colourings of G_1 and G_2 together we obtain a 4-colouring of G. This is easy if the marked vertex is the same in the two colourings. Otherwise we need to use Kempe-chain arguments to change one or both of the colourings. There are two cases: either the two marked vertices are adjacent in the cycle C, or they are not.

Case 1. We deal first with the adjacent case. Suppose that the colouring of C in G_1' is (r, g, b, g, b) and the colouring of C in G_2' is (b, r, g, b, g), where the vertices are labelled in order v_1, v_2, v_3, v_4, v_5 (see Fig. 10.7). Thus, the marked vertex of C in G_1' is v_1, and in G_2' is v_2. If there is no blue–red chain in G_2 from v_2 to v_4, then we can change the G_2-colouring of C to (r, b, g, b, g), and now the marked vertices match up. Thus, by relabelling the colours in either G_1 or G_2, we obtain a 4-colouring of G.

Otherwise, there is a blue–red chain in G_2 from v_2 to v_4, so there cannot be a green–yellow chain from v_3 to v_5, and therefore we can change the G_2-colouring to (b, r, y, b, g). Now we consider a completely different method of obtaining a colouring for G_1. This time we do not add an extra vertex, we simply collapse the cycle C by identifying v_1 with v_4. This gives us a new graph G_1'', which is again a triangulation containing fewer vertices than G. A four-colouring of G_1'' now induces a colouring of G_1 in which v_1 and v_4 are coloured with the same colour. Therefore, the G_1-colouring of C is, without loss of generality, of the form (b, r, y, b, x), where x is a colour other than blue.

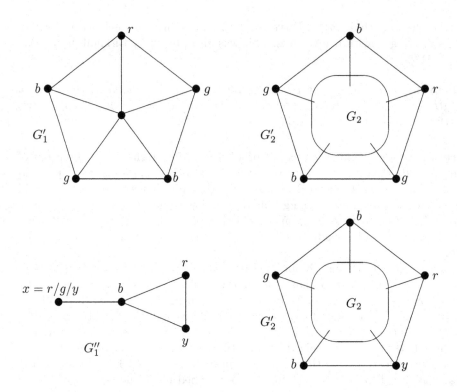

Fig. 10.7 Case 1 of Theorem 10.7.

If x is green, the two colourings of C already match up, and we are done. If x is yellow, we obtain a 3-colouring of C in which the marked vertex is v_2. Therefore (changing the names of the colours suitably), we obtain a match with the original G_2 colouring (b, r, g, b, g).

This leaves just the case when x is red, so that the G_1-colouring of C is (b, r, y, b, r) and the marked vertex is v_3. So in effect the marked vertex in the G_1-colouring has jumped (from v_1 to v_3) over the marked vertex (v_2) in the G_2-colouring. Repeating the argument, with the roles of G_1 and G_2 interchanged, we can obtain a G_2-colouring with marked vertex v_4 and a G_1-colouring with marked vertex v_3. Repeating the argument twice more, both marked vertices move on two more places around the cycle, to v_1 and v_5, respectively. In particular, we have obtained a colouring of G_2 with marked vertex v_1. Therefore, we can match this up with the original G_1-colouring (r, g, b, g, b), and we are done.

Case 2. The other case is the non-adjacent case, where we can suppose that the G_1-colouring of C is (r, g, b, g, b), with marked vertex therefore v_1, and the G_2-colouring is (g, b, r, g, b), with marked vertex v_3 (see Fig. 10.8). Again, we consider red–blue Kempe chains. If there is no such chain from v_2 to v_5 in G_2, then we can recolour G_2 in such a way that C is coloured (g, r, b, g, b). Now the marked vertex is v_2, and we are back in the first case.

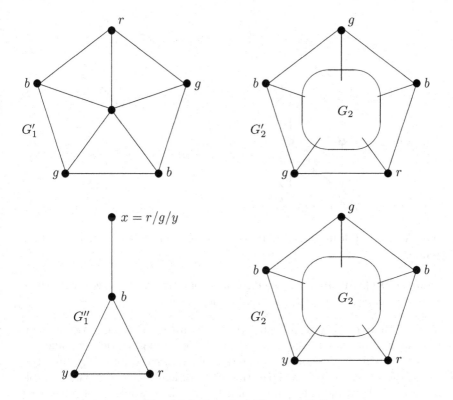

Fig. 10.8 Case 2 of Theorem 10.7.

Otherwise, there is a red–blue chain in G_2 from v_2 to v_5, so there is no green–yellow chain from v_1 to v_4. Therefore, we can change the G_2-colouring of C to (g, b, r, y, b). Now we collapse G_1 again by identifying v_2 and v_5, to get a colouring of G_1 in which C is coloured (x, b, r, y, b), for some colour x other than blue.

If x is green we match the new G_2-colouring, and we are done. If x is yellow we rename the colours y and g to match the original G_2-colouring. Finally, if x is red, the marked vertex in the G_1-colouring is v_4, adjacent to the marked vertex in the orignal G_2-colouring, and we are back in the first case again. □

Definition 10.8 *A triangulated graph with the property that the only separating cycles of length at most 5 are those which disconnect a single vertex, is called an* **internally 6-connected triangulation.**

10.4 Larger configurations

Nearly 10 years after Birkhoff's proof of reducibility of the Birkhoff diamond, his student Franklin proved by a similar argument the reducibility of the configuration consisting of a vertex of degree 6 with three consecutive neighbours of degree 5 (see Fig. 10.9). Franklin's paper, which contains much else besides, is reproduced in full by Biggs, Lloyd and Wilson [10].

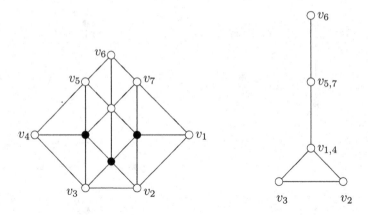

Fig. 10.9 A vertex of degree 6 with three consecutive neighbours of degree 5, and the collapsed heptagon.

Theorem 10.9 (Franklin) *The configuration of a vertex of degree 6 with three consecutive neighbours of degree 5 is reducible.*

Proof This time, the key to reducing the configuration is to collapse the heptagon by identifying vertices v_5 and v_7, and identifying vertices v_1 and v_4. First note that this is possible—if v_5 and v_7 were already joined by an edge, then the graph would contain a separating cycle of length 3, contrary to Corollary 10.6. Similarly, if v_1 and v_4 were joined, then there would be a separating cycle of length 5, containing more than one vertex in its interior, contrary to Theorem 10.7.

There are only five essentially different colourings of the collapsed heptagon, as shown in Fig. 10.10. The first three of these can be immediately extended to a colouring inside the heptagon, as shown in Fig. 10.11(a)–(c). In the other two cases, we use a Kempe-chain argument as before.

In case 4, there cannot be both a blue–green chain from v_5 to v_2, and a red–yellow chain from v_1 to v_3. So, we can either change the colour of v_2 from green to blue, giving case 3 again, or change the colour of v_1 to yellow, and complete the colouring as in Fig. 10.11(d).

In case 5, there cannot be both a yellow–blue chain from v_2 to v_5 and a red–green chain from v_4 to v_6. So we can either change the colour of v_2 to yellow, giving case 2, or change the colour of v_6 to red, giving case 3 again. This completes the proof of Franklin's theorem. □

In what we have done so far, we have tacitly assumed that the vertices of the boundary ring of our configuration are distinct. In the cases we have considered, this is easy to prove, for otherwise there is a separating cycle of length 4 or less, contrary to Theorem 10.6. Indeed, provided our configuration has all its vertices at distance 2 or less from a particular vertex, then all the vertices of the boundary ring are distinct by Theorem 10.7.

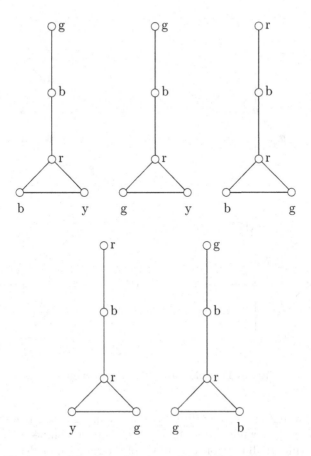

Fig. 10.10 The five colourings of the collapsed heptagon.

However, for larger configurations this is not necessarily true. This extra complication is known as the **immersion problem**. The significance of it is that our Kempe-chain arguments do not work if the boundary ring is not a cycle. Thus extra reduction arguments are needed in this case.

10.5   Using a computer to prove reducibility

The above examples give the flavour of the reducibility arguments used in the proof of the four-colour theorem. Once we have produced an unavoidable set of configurations, we need to try to prove reducibility for each configuration in the set. Following Kempe's method, we remove the given configuration from the graph, leaving just its ring of neighbours behind. Then by induction we can 4-colour the resulting graph, as it has fewer vertices than the original. The problem now is to extend the colouring to the configuration inside the ring. In principle, we need to consider all possible colourings for the ring of neighbours of the configuration, and produce a colouring of the whole graph in each case. Some

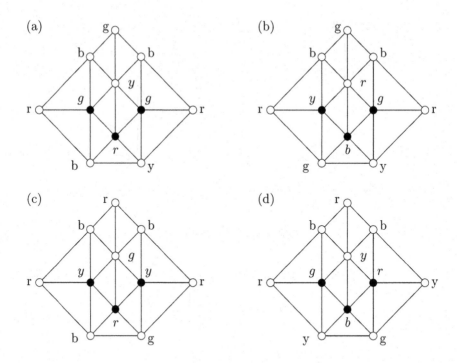

Fig. 10.11 Colouring inside the heptagon.

colourings of the ring will extend directly to a colouring inside (these are called **good** ring-colourings), while others may succumb to a Kempe-chain argument to recolour the ring. If all possible colourings of the ring can be dealt with in this way, Appel and Haken, following Heesch, called the configuration **D-reducible**. Thus the D-reducible configurations are exactly the ones which can be dealt with by the original methods of Kempe.

Now it is easy to program a computer to enumerate all possible colourings of the ring, and it is easy to program it to check for each ring-colouring whether there is a way to extend it to a colouring of the whole configuration. But what about the Kempe-chain arguments? Well, there are just three choices for the partition of the four colours into three pairs, and for each colouring and each partition, there are only finitely many possibilities for incompatible pairs of Kempe chains. In each case, there are just two possible colour changes to try. Thus it is straightforward also to program a computer to apply all Kempe-chain arguments systematically, to see if we can reduce each possible ring colouring to a good colouring. Note that this may require several successive recolourings, but there is a bound on the number of recolourings needed. In this way, we can write a straightforward program to check whether a given configuration is D-reducible.

As we have seen in Sections 10.1 and 10.4, a more powerful method of proving reducibility is to replace the interior of the ring by something smaller. The most obvious way to do this is simply to triangulate inside the ring in any way we like. This will reduce the number of possible colourings of the ring, by making some vertices adjacent to others, and may, if we are lucky, eliminate the awkward ones.

More subtle ways to do this include identifying two or more vertices of the ring, so that they must be coloured the same colour, or introducing new vertices (as long as they are fewer than in the original configuration). This will again produce a 4-colourable graph, but the possible colourings of the ring will be more restricted. The idea is to choose a suitable replacement for the configuration, to exclude the awkward cases for the ring colouring. If this works, the configuration is called **C-reducible**. This is less easy to program for a computer, as there may be a huge number of ways to replace the interior of the ring with something smaller, and it is not clear beforehand which ones are likely to be useful. It is out of the question to try all possibilities. So in practice, Heesch, and later Appel and Haken, just tried a few likely-looking candidates, and if these did not work, they gave up.

Exercises

Exercise 10.1 Use the Theorems of Birkhoff and Franklin to show that if a vertex of degree 5 has three neighbours of degree 5, one of degree 6, and one of arbitrary degree, then it is reducible.

Exercise 10.2 Consider the configuration of a vertex of degree 5 with two consecutive neighbours of degree 5, and the other neighbours of degree 6. By collapsing this down to a 4-star as in Fig. 10.12, show that this configuration is reducible.

Exercise 10.3 Use Exercise 10.2 to show that a vertex of degree 5 with two neighbours of degree 5 and three neighbours of degree 6 is reducible.

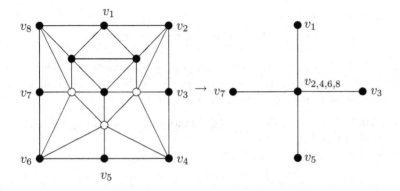

Fig. 10.12 The configuration for Exercise 10.2.

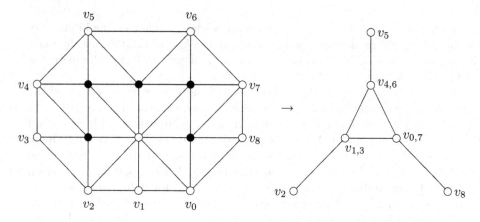

Fig. 10.13 The configuration for Exercise 10.4.

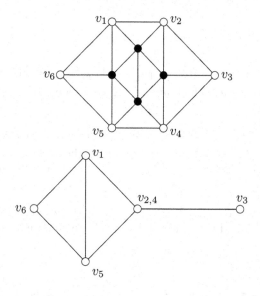

Fig. 10.14 Diagram for Exercise 10.5.

Exercise 10.4 (Hard) Consider the configuration of a vertex of degree 8 with five consecutive neighbours of degree 5. By collapsing this configuration as shown in Fig. 10.13, show that this configuration is reducible.

Exercise 10.5 Prove that the Birkhoff diamond is reducible by collapsing the hexagon as shown in Fig. 10.14.

11
Discharging

11.1 Unavoidable sets

In order to prove the four-colour theorem, we need a good method for producing unavoidable sets. All methods ultimately rely on Euler's formula, or more specifically on the fact that for triangulated graphs (i.e. maximal planar graphs)

$$\sum_{\text{all vertices } v} (6 - d(v)) = 12,$$

which is Proposition 3.18. For completeness, we prove this here.

Lemma 11.1 *In a triangulated graph or multigraph, $\sum_v (6 - d(v)) = 12$.*

Proof The faces have exactly three edges, so $2q = 3r$, and thus $6r - 4q = 0$. Also, $\sum_v d(v) = 2q$ and $\sum_v 6 = 6p$, so

$$\sum_v (6 - d(v)) = 6p - 2q$$

$$= 6p - 2q + (6r - 4q)$$

$$= 6(p - q + r) = 12$$

by Euler's formula. $\qquad\qquad\square$

The main difficulty in using this result is that it is a **global** result (i.e. it mentions **all** vertices) and we want **local** consequences (i.e. the existence of certain small configurations).

Let us look at some of the early results on unavoidable sets, in order to appreciate where the difficulty lies. The earliest significant result after Kempe's was that of Wernicke, who strengthened Kempe's result by proving that some vertex of degree 5 must have a neighbour of degree 5 or 6. (In fact, he proved this in the dual form, for maps rather than graphs.)

First, we introduce a shorthand notation for an unavoidable set $\mathcal{U}$ of configurations. We always assume that $\mathcal{U}$ contains the first three configurations of Fig. 9.1 (i.e. a vertex of degree 2, 3, or 4), so we do not explicitly mention them each time. In effect, this means that from now on we may assume that our graphs never have any vertices of degree less than 5. A vertex of degree 5 is denoted by ●, and a vertex of degree 6 by ○. Vertices of larger degree are denoted by circles labelled with the degree. Thus, Kempe's unavoidable set in Fig. 9.1 is denoted by {●}. Similarly, Wernicke's unavoidable set is denoted by {●——●, ●——○}.

Theorem 11.2 (Wernicke, 1904) *In any minimal counterexample to the (vertex-colouring) four-colour theorem, there is a vertex of degree 5 with a neighbour of degree 5 or 6.*

Proof We first choose any planar embedding of the graph, and note that all faces are triangles. The idea now is to count the faces (triangles) of the graph which are incident with a vertex of degree 5 or 6. If no vertex of degree 5 is adjacent to any vertex of degree 5 or 6, then each vertex of degree 5 is surrounded by five such faces, and each such face is incident with a unique vertex of degree 5. Thus, every vertex of degree 5 contributes 5 to the number of such faces. A vertex of degree 6 is surrounded by six such faces, but each such face could be incident with three vertices of degree 6, so could be counted up to three times. So each vertex of degree 6 contributes at least two faces (on average). Therefore, the number of faces, r, is at least $5p_5 + 2p_6$, where p_i denotes the number of vertices of degree i. Thus,

$$\begin{aligned} r &\geqslant 5p_5 + 2p_6 \\ &\geqslant 5p_5 + 2p_6 - p_7 - 4p_8 - \cdots \\ &= \sum_{i=5}^{\infty}(20 - 3i)p_i \\ &= 20p - 3\sum_{i=5}^{\infty} ip_i. \end{aligned}$$

Now the handshaking lemma implies

$$\sum_{i=5}^{\infty} ip_i = 2q = 3r$$

$$\Rightarrow 20q - 21r = 10\sum_{i=5}^{\infty} ip_i - 7\sum_{i=5}^{\infty} ip_i$$

$$= 3\sum_{i=5}^{\infty} ip_i.$$

Substituting back into the above inequality gives

$$\begin{aligned} r &\geqslant 20p - 20q + 21r \\ &= r + 40 \end{aligned}$$

by Euler's formula, giving a contradiction. □

Nearly 20 years after Wernicke, Franklin strengthened Wernicke's result, and proved that some vertex of degree 5 must be adjacent to **two** vertices of degree 5 or 6. The proof he gives is very similar. Both results are proved again in 1940, in a paper by Lebesgue [35], in which he generalizes the method to produce a variety of different unavoidable sets. These and other results along these lines are discussed in detail in Ore's book [37].

Lebesgue's version of the proof takes Euler's formula in the form

$$2\sum_{i=5}^{\infty} p_i - 2q + 2r = 4$$

and uses the handshaking lemmas to substitute for $2q$ as either $\sum_{i=5}^{\infty} ip_i$ or $3r$, giving the two equations

$$\sum_{i=5}^{\infty} (2 - i)p_i + 2r = 4$$

$$\sum_{i=5}^{\infty} 2p_i - r = 4.$$

He then eliminates p_7 from these two simultaneous equations to give

$$4p_5 + 2p_6 - 2p_8 - \cdots = 28 + r$$

which implies $r < 4p_5 + 2p_6$, contradicting $r \geqslant 5p_5 + 2p_6$. He then goes on to note that there is enough room between these two inequalities to prove Franklin's result as well.

Theorem 11.3 (Franklin, 1923) *In any minimal counterexample to the four-colour theorem, there is a vertex of degree 5 with two neighbours, each of degree 5 or 6.*

Proof We summarize Lebesgue's proof of this result. As above, we see that $r < 4p_5 + 2p_6$. Now suppose that no vertex of degree 5 is adjacent to two vertices of degree 5 or 6. Then, each vertex of degree 6 contributes at least 2 to the number of triangles, as before. Two adjacent vertices of degree 5 contribute 8 triangles between them, that is, an average of four each. A vertex of degree 5 adjacent to a vertex of degree 6 contributes 5 triangles, less the amount $\frac{2}{3}$ which we have already counted of the 2 triangles incident to both vertices. An isolated vertex of degree 5 contributes 5. In each case, therefore, the vertex of degree 5 contributes at least 4, so in total the number of triangles is at least $4p_5 + 2p_6$. Thus, we obtain $r \geqslant 4p_5 + 2p_6$, and the required contradiction. □

Note that the unavoidable set in this case is

not as stated in [8, vol. 1, chapter 4].

11.2 Simple examples of discharging

At this point, it becomes clear that it is necessary to consider not just the neighbours of a vertex of degree 5, but the neighbours of the neighbours, and possibly more. Everything becomes very complicated, and in particular, the counting arguments get out of hand. The idea of 'discharging' was introduced by Heesch to overcome this difficulty. The word itself is due to Haken, as Heesch used the word 'curvature' instead of 'charge', but the concept is the same.

(Heesch's work on the four-colour theorem began in the late 1940s, but for various reasons was not published until 1969, which may account in part for the lack of recognition he has been given for this work. In particular, Ore did not mention the work of Heesch in his 1967 book on the four-colour problem [37], which was considered the authoritative work on the subject at that time. Nevertheless, Haken learnt from Heesch many of the ideas which played a part in the eventual proof. The introduction to the Appel–Haken [4] paper is instructive in this regard.)

We put a 'charge' of $6 - d(v)$ on each vertex v. This means that the vertices of degree 5 receive a charge of $+1$, the vertices of degree 6 are uncharged, and all other vertices have a negative charge. Thus by Lemma 11.1, the total charge on the vertices of the graph is 12. Then we redistribute the charge in some way, and try to get all charges to be 0 or negative. This is obviously impossible, by charge conservation, so we must see what 'obstacles' there are. Since every graph must contain some obstacle preventing complete discharging, we obtain some set of configurations with the property that every graph contains at least one of them. This set is called an **unavoidable set**. Ultimately, our goal is to produce an unavoidable set such that we can show that every configuration in the unavoidable set is reducible.

(Roughly speaking, if discharging works everywhere, locally, then it works globally. But this contradicts Euler's formula, so there must be somewhere where discharging fails. We then look to see what local conditions are necessary for discharging to fail.)

To try to understand the method, let us first obtain Kempe's unavoidable set by the discharging method. In this case, we use the simplest possible discharging algorithm, that is, the algorithm that does nothing at all to the original charges. Then, the only vertices which are not 'discharged' (by which we mean the charge on the vertex is zero or negative) are those of degree at most 5. These vertices are then the obstacles which constitute the unavoidable set.

We look now at a less trivial example, namely Wernicke's theorem. Consider the following discharging algorithm.

Algorithm 11.4 *Every vertex of degree 5 gives a charge of $\frac{1}{5}$ to each of its neighbours which has degree at least 7.*

Now it is clear that this will not discharge a vertex of degree 5 unless all its neighbours have degree at least 7. Therefore, ●——● and ●——○ must appear in the unavoidable set. What we now show is that these are the **only** obstructions to complete discharging. That is, we show that

$$\mathcal{U}_1 = \{\,\bullet\!\!-\!\!\bullet\,,\,\bullet\!\!-\!\!\circ\,\}$$

is an unavoidable set, thereby giving an alternative proof of Theorem 11.2. A consequence of this is that in trying to prove the four-colour theorem, we may assume that our vertex of degree 5 has a neighbour of degree 5 or 6.

Theorem 11.5 *The set $\mathcal{U}_1$ defined above is an unavoidable set.*

Proof Now if G has none of the above configurations in $\mathcal{U}_1$, then all neighbours of every vertex of degree 5 have degree at least 7, so the vertices of degree 5 end up with a charge of 0. The vertices of degree 6 are not affected by the algorithm, so end up with the 0 charge they started with. If v is a vertex of degree $k \geqslant 7$, then no two consecutive neighbours of v can have degree 5, so v has at most $\frac{1}{2}k$ neighbours of degree 5, so acquires a charge of at most $\frac{1}{10}k$. Since it started out with a charge of $6-k$, it ends up with a charge of at most $6 - \frac{9}{10}k \leqslant 6 - \frac{9}{10}\cdot 7 < 0$. Thus the total charge on G is negative. This contradiction implies that $\mathcal{U}_1$ is an unavoidable set. $\qquad\square$

Notice that a single discharging algorithm can be used to prove unavoidability for different sets of configurations. It all depends on how, and how closely, we look at the graph. For example, consider the following algorithm.

Algorithm 11.6 *Each vertex of degree 5 gives a charge of $\frac{1}{3}$ to each of its neighbours of degree 7 or more.*

Algorithm 11.6 can be used to prove that $\mathcal{U}_1$ defined above is an unavoidable set, by a slight modification of the above argument (see Exercise 11.1). More usefully, it can be used to show that

$$\mathcal{U}_2 = \left\{\,\bullet\!\!-\!\!\bullet\,,\,\triangle\,\right\}$$

is an unavoidable set.

Theorem 11.7 *The set $\mathcal{U}_2$ defined above is an unavoidable set.*

Proof Suppose that G is a triangulated graph with minimal vertex degree 5, and suppose that G contains none of the configurations in $\mathcal{U}_2$. Thus, no vertex of degree 5 has either a neighbour of degree 5, or two consecutive neighbours of degree 6. So each vertex v of degree 5 has at least three neighbours of degree 7 (for otherwise it has at most two such, so it has two consecutive neighbours of degree at most 6—either both have degree 6, or one has degree 5). Therefore, the vertices of degree 5 are discharged. As before, vertices of degree 7 have at most three neighbours of degree 5, so end up with a charge at most $(6-7) + 3 \cdot \frac{1}{3} = 0$, while vertices of degree $k \geqslant 8$ have charge at most $(6-k) + \frac{1}{6}k = 6 - \frac{5}{6}k \leqslant 6 - \frac{5}{6}\cdot 8 < 0$. $\qquad\square$

In a similar way, Appel and Haken used the same discharging procedure to produce their unavoidable set of 1834 configurations, as to produce their later set of 1482 configurations.

11.3 A more complicated discharging algorithm

In this section, we illustrate one of the ways in which a more subtle discharging algorithm can be used to produce a different sort of unavoidable set. We show that the following set $\mathcal{U}_3$ is an unavoidable set. This example is given by Saaty and Kainen [44], following Haken [27].

$$\mathcal{U}_3 = \left\{ \circ, \textcircled{7}, \text{◆}, \text{▨}_8, \text{▨}_9, \text{▨}_{10}, \text{▨}_{11} \right\}.$$

We have already seen that the configuration of a vertex of degree 5 with three consecutive neighbours of degree 5 (the so-called Birkhoff diamond) is reducible. The vertex of degree 8 with five consecutive neighbours of degree 5 was proved irreducible by Choinacki [18] in 1942 (see also Exercise 10.4, and [23, Theorem 6.5.5] for a sketch). The last three configurations in the set $\mathcal{U}_3$ were proved irreducible by Franklin [22] in 1922. As a corollary, we see that every minimal counterexample to the four-colour theorem must contain a vertex of degree 6 or 7.

Consider the following discharging algorithm.

Algorithm 11.8 *For each vertex of degree at least 8, we distribute its charge among those of its neighbours which have degree 5, according to a certain weighting. We consider chains of consecutive neighbours of degree 5, and weight ends of chains with 1, and interior vertices of chains with 2, and also isolated vertices 2. The charge is then distributed so that the vertices of weight 2 get twice the charge that those of weight 1 get.*

We illustrate this algorithm by means of an example. In Fig. 11.1, we draw the neighbourhood of a particular vertex of degree 9. Vertices of degree 5 are

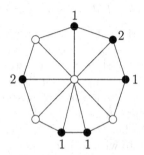

Fig. 11.1 Weights on the neighbours of a vertex of degree at least 8.

denoted by black circles, and those of degree 8 or more by white circles. The weights of the vertices of degree 5 are marked on the figure. The total weight is 8, so the vertices of weight 1 receive $\frac{1}{8}$ of the original charge of the central vertex, while the vertices of weight 2 receive $\frac{1}{4}$. Since the central vertex has degree 9 in this case, it has charge -3, so its neighbours receive charges $-\frac{3}{8}$ or $-\frac{3}{4}$ from it, according to their weight.

Thus, by construction, the algorithm discharges all vertices of degree at least 8. We need only show that, provided none of the configurations in $\mathcal{U}_3$ occurs, it discharges the vertices of degree 5. This contradiction then shows that $\mathcal{U}_3$ is an unavoidable set of configurations.

We prove a technical lemma first.

Lemma 11.9 *If G is a plane graph in which none of the configurations in $\mathcal{U}_3$ occurs, then the amount of charge moved by Algorithm 11.8 from a vertex v of degree j to a vertex u of degree 5, is at most $-w/4$, where w is the weight of u as a neighbour of v.*

Proof Note that all the charges on vertices of degree greater than 6 are **negative**, so check carefully the directions of the inequalities below. If $j \geqslant 12$, then $6 - j \leqslant -j/2$, and as the total weight W of the neighbours of v is at most $2j$, the transferred charge is

$$\frac{w}{W}(6 - j) \leqslant \frac{w}{2j}(6 - j)$$

$$\leqslant \frac{w(-j/2)}{2j}$$

$$= -\frac{w}{4}.$$

If $j = 9, 10$, or 11, then by assumption, v has at most $j - 2$ neighbours of degree 5. The number of weight 1 neighbours of v is even. If this number is 0, then we have only isolated points (here we use the fact that not all neighbours have degree 5), which make up at most half of the neighbours, so the total weight is at most j, giving transferred charge

$$\frac{w}{W}(6 - j) \leqslant \frac{w}{j}(6 - j)$$

$$\leqslant -\frac{w}{3},$$

since

$$j \geqslant 9 \Rightarrow 6 - \frac{2}{3}j \leqslant 0$$

$$\Rightarrow 6 - j \leqslant -\frac{1}{3}j$$

$$\Rightarrow \frac{6 - j}{j} \leqslant -\frac{1}{3}.$$

If on the other hand there are at least two neighbours of weight 1, then the total weight is at most $2j - 6$ (here, we use the fact that at least two neighbours do not have degree 5), which gives the result again in the same way:

$$\frac{w}{W}(6 - j) \leqslant \frac{w}{2j - 6}(6 - j)$$

$$\leqslant -\frac{w}{4}$$

since

$$9 - j \leqslant 0 \Rightarrow 12 - 2j \leqslant 3 - j$$

$$\Rightarrow 6 - j \leqslant -\frac{2j - 6}{4}$$

$$\Rightarrow \frac{6 - j}{2j - 6} \leqslant -\frac{1}{4}.$$

Finally, if $j = 8$, we apply a similar argument. Since we are assuming that the vertex of degree 8 does not have five consecutive neighbours of degree 5, it follows that all the chains of such neighbours have length at most 4. Then, it is easy to draw all possible cases, and see that the lengths of chains can be $\{4, 2\}$, $\{4, 1\}$, $\{3, 3\}$, $\{3, 2\}$, $\{3, 1, 1\}$, $\{2, 2, 1\}$, $\{2, 1, 1\}$, $\{1, 1, 1, 1\}$, or subsets of these (see Fig. 11.2). In all these cases, the total weight is 8 or less, and therefore, the charge transferred is at most $(6 - 8)w/8 = -w/4$, as required. □

Theorem 11.10 *If G is a triangulated graph with minimum vertex degree 5, and G contains none of the configurations in $\mathcal{U}_3$, then Algorithm 11.8 discharges G.*

Proof We pick a vertex u of degree 5, and divide the proof into three cases, according as the number of neighbours of v degree 5 is 3, 2 or at most 1.

In the case when u has three neighbours of degree 5, we see that u has weight 2 in the shareouts of both its other two neighbours, v_1 and v_2 (see Fig. 11.3, where the weight of u as a neighbour of v_i is shown alongside the edge uv_i) so by Lemma 11.9 gets a charge of at most -1. Therefore, the final charge is at most 0.

In the case when u has two neighbours of degree 5, it has weights 2, 1 and 1 in the three shareouts so again by Lemma 11.9 the final charge is at most 0.

Finally, if u has 1 or 0 neighbours, the weights are 2, 2, 1, 1 in the first case, or 2, 2, 2, 2, 2 in the second, and again the result follows. □

Corollary 11.11 *The set $\mathcal{U}_3$ of configurations, defined above, is an unavoidable set, for a minimal counterexample to the four-colour conjecture.*

Combining this with known reducibility results, we have the following.

Corollary 11.12 *Any minimal counterexample to the four-colour theorem contains a vertex of degree 6 or 7.*

11.4 Conclusion

To prove the four-colour theorem, therefore, it suffices to devise a sufficiently complicated discharging algorithm, and a sufficiently complicated 'unavoidable

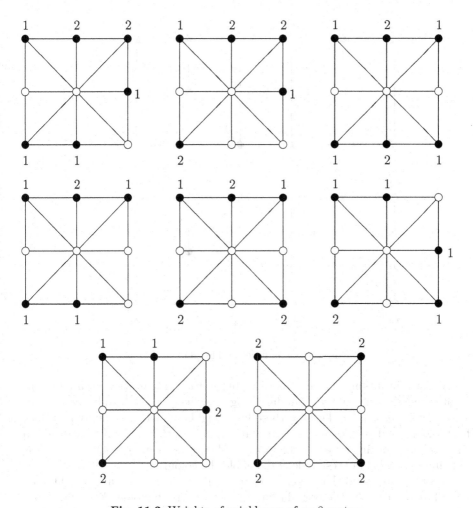

Fig. 11.2 Weights of neighbours of an 8-vertex.

set', and then prove that (a) the discharging algorithm will discharge any graph that avoids the unavoidable set, and (b) any graph in the unavoidable set is reducible.

The difficulty, of course, is to achieve both these aims simultaneously. After running his reducibility-testing program on very many configurations, Heesch developed a good intuition as to which configurations were likely to be easily proved to be reducible, and which were likely to be troublesome. Only then did he start serious work on discharging procedures. (Note: Heesch actually called the charge 'curvature', on the basis that you need a fixed amount of curvature in your map in order to roll it up into a sphere. You can move the curvature around, as long as you keep the total fixed. It was Haken who renamed the concept 'charge' by analogy with electrical networks.)

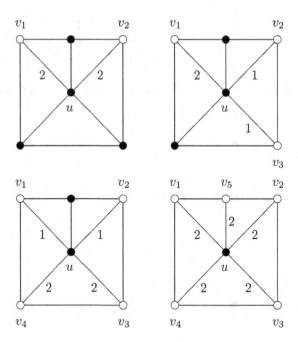

Fig. 11.3 The proof of Theorem 11.10.

Heesch's intuition told him that a vertex in the configuration with four consecutive neighbours in the boundary ring was likely to be troublesome. In a sense this is reasonable, because these four neighbours can be coloured with all four colours, and even with a Kempe-chain argument, this can only be reduced to three colours, leaving a unique colour available for the given interior vertex—this may impose too many restrictions to enable the colouring to be completed.

Similarly, he found that a vertex in the configuration with three neighbours, not all consecutive, in the boundary ring, was also troublesome. A configuration without either of these two obstacles was called **geographically good**. A third obstacle was a pair of adjacent vertices v, w of degree 5, each adjacent to only one other vertex inside the boundary ring (the same vertex for both v and w).

The strategy of Appel and Haken at this point was to start with a simple discharging algorithm (distributing the positive charge of each vertex of degree 5 equally to all its neighbours of degree at least 7), and then to modify it repeatedly, to exclude troublesome cases at each stage. Hundreds of modifications were needed until a reasonable unavoidable set was produced—first of geographically good configurations, and then with more modifications, also excluding the third type of obstacle.

Appel and Haken [4] used a discharging algorithm made up of over 300 separate rules, whereas Robertson *et al.* [42] found a simpler algorithm of just 32 rules.

So can we regard the four-colour theorem as finally proved? This was the first major theorem whose proof involved a substantial amount of computer calculation, and as such it was bound to cause controversy. It has stimulated a wide-ranging debate over the past quarter of a century, into what we mean by a proof, what a proof is for, and so on. Whilst some traditionalists still insist that a proof is not a proof unless it is written out on paper, the modern view is that all means of proof are fair, provided adequate attention is paid to verifiability and reproducibility.

In both respects, we have come a long way since 1976. The almost universal availability of high-powered computers, running high-quality mathematical software, means that most computer-assisted proofs nowadays are easy to verify, and easy to reproduce. Of course, the harder proofs are not so easy to verify and to reproduce, but that is also the case with proofs 'by hand', and must be expected.

It is noticeable, too, that in modern mathematics the importance of proof in the traditional sense is declining. In certain areas, the emphasis has shifted to getting the 'right answer'. In very many cases, too, answers can be obtained by computer to problems that could never be solved by hand. The purpose of a proof is then to demonstrate that this really is the right answer. But can you trust a 'proof', written out on paper, that consists of 10,000 pages of intricate argument? Such 'proofs' certainly exist, but you can no more check them yourself than you can check a computer calculation line by line. Thus many mathematicians nowadays are looking for better ways than traditional proofs, of convincing ourselves, and each other, that the answer is right.

The fact is, that the mathematical literature is riddled with false 'proofs', like that of Kempe, and this is a problem which will not go away. How are we to know which of today's generally accepted proofs will turn out in ten years time to be fallacious?

Exercises

Exercise 11.1 Use Algorithm 11.6 to prove that

$$\mathcal{U}_1 = \{\bullet\!\!-\!\!\bullet\,,\,\bullet\!\!-\!\!\circ\}$$

is an unavoidable set.

Exercise 11.2 Consider the discharging algorithm which distributes the charges on each vertex of degree 5 equally to all its neighbours of degree at least 9.

Use this algorithm to show that

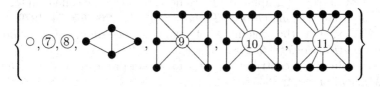

is an unavoidable set.

Bibliography

[1] K. Appel, The proof of the four-colour theorem, *New Scientist*, **72**: 154–5, 1976.

[2] K. Appel and W. Haken, The solution of the four-color-map problem, *Scientific American*, **237**(4): 108–21, 1977.

[3] K. Appel and W. Haken, Every planar map is four colorable, *Contemporary Mathematics*, Vol. 98. American Mathematical Society, 1989.

[4] K. Appel and W. Haken, Every planar map is four colourable, Part I: discharging, *Illinois Journal of Mathematics*, **21**: 429–90, 1977.

[5] K. Appel, W. Haken and J. Koch, Every planar map is four colourable, Part II: reducibility, *Illinois Journal of Mathematics*, **21**: 491–567, 1977.

[6] W. W. R. Ball, *Mathematical Recreations and Essays*, 1892 and many subsequent editions.

[7] D. Barnette, *Map Coloring, Polyhedra, and the Four-color Problem*, Math. Assoc. of America, 1983.

[8] L. W. Beineke and R. J. Wilson, *Selected Topics in Graph Theory, 1, 2, 3*, Academic Press, 1978, 1983 and 1988.

[9] C. Berge, *Graphs*, 3rd ed., North-Holland, 1985.

[10] N. L. Biggs, E. K. Lloyd and R. J. Wilson, *Graph Theory 1736–1936*, Oxford University Press, 1976.

[11] G. D. Birkhoff, A determinant formula for the number of ways of coloring a map, *Annals of Mathematics (2)*, **14**: 42–46, 1912-3.

[12] G. D. Birkhoff, The reducibility of maps, *American Journal of Mathematics*, **35**: 115–28, 1913.

[13] B. Bollobás, *Graph Theory—An Introductory Course*, Graduate Texts in Mathematics, Springer-Verlag, 1979.

[14] J. A. Bondy and U. S. R. Murty, *Graph Theory with Applications*, Macmillan, 1976.

[15] R. L. Brooks, On colouring the nodes of a network, *Proceedings of the Cambridge Philosophical Society* **37**: 194–7, 1941.

[16] P. A. Catlin, Hajós' graph-coloring conjecture: variations and counterexamples, *Journal of Combinatorial Theory (Series B)*, **26**: 268–74, 1979.

[17] A. Cayley, On the colouring of maps, *Proceedings of the London Mathematical Society*, **9**: 148, 1878.

[18] C. A. Choinacki, A contribution to the four color problem, *American Journal of Mathematics*, **64**: 36–54, 1942.

[19] R. Diestel, *Graph Theory*, Springer, 1997. 2nd ed., 2000.

[20] G. A. Dirac, A property of 4-chromatic graphs and some remarks on critical graphs, *Journal of the London Mathematical Society (Series 1)*, **27**: 85–92, 1952.

[21] S. Fiorini and R. J. Wilson, *Edge-colourings of Graphs*, Research Notes in Mathematics, Vol. 16, Pitman, 1977.

[22] P. Franklin, The four color problem, *American Journal of Mathematics*, **44**: 225–36, 1922.

[23] R. Fritsch and G. Fritsch, *The Four-color Theorem: History, Topological Foundations, and Idea of Proof*, Springer, 1998 (translation of the German original).

[24] J. L. Gross and T. W. Tucker, *Topological Graph Theory*, Wiley, 1987.

[25] H. Hadwiger, Über eine Klassifikation der Streckenkomplexe, *Vierteljahrschriften der Naturforschungsgesellschaft Zürich*, **88**: 133–142, 1943.

[26] G. Hajós, Über eine Konstruktion nicht *n*-färbbarer Graphen, *Wissenschaftliche Zeitschrift der Martin-Luther-Universität Halle-Wittenberg. Mathematisch-Naturwissenschaftliche Reihe*, **10**: 116–17, 1961.

[27] W. Haken, An existence theorem for planar maps, *Journal of Combinatorial Theory (Series B)*, **14**: 180–4, 1973.

[28] F. Harary, *Graph Theory*, Addison-Wesley, 1969.

[29] T. R. Jensen and B. Toft, *Graph Coloring Problems*, Wiley, 1995.

[30] A. B. Kempe, On the geographical problem of the four colours, *American Journal of Mathematics*, **2**: 193–200, 1879.

[31] T. P. Kirkman, On the representation of polyhedra, *Philosophical Transactions of the Royal Society of London*, **146**: 413–18, 1856.

[32] T. P. Kirkman, On the properties of the R-pyramid, being the first class of R-gonous X-edra, *Philosophical Transactions of the Royal Society of London*, **148**: 145–61, 1858.

[33] D. König, *Theorie der endlichen und unendlichen Graphen* (1936), translated as *The theory of Finite and Infinite Graphs*, Birkhäuser, 1990.

[34] K. Kuratowski, Sur le problème des courbes gauche en topologie, *Fundamenta Mathematicae*, **15**: 271–83, 1930.

[35] H. Lebesgue, Quelques conséquences simples de la formule d'Euler, *Journal de Mathématiques Pures et Appliquées, Neuvième Série*, **19**: 27–43, 1940.

[36] K. O. May, The origin of the four-colour conjecture, *Isis*, **56**: 346–8, 1965.

[37] O. Ore, *The Four-color Problem*, Academic Press, 1967.

[38] J. Petersen, Sur le théorème de Tait, *L'Intermédiaire des Mathématiciens*, **5**: 225–7, 1898.

[39] R. C. Read, Introduction to chromatic polynomials, *Journal of Combinatorial Theory*, **4**: 52–71, 1968.

[40] G. Ringel, *Map Color Theorem*, Springer, 1974.

[41] N. Robertson, P. Seymour and R. Thomas, Hadwiger's conjecture for K_6-free graphs, *Combinatorica*, **13**: 279–361, 1993.

[42] N. Robertson, D. P. Sanders, P. Seymour and R. Thomas, The four-colour theorem, *Journal of Combinatorial Theory (Series B)*, **70**: 2–44, 1997.

[43] N. Robertson, D. P. Sanders, P. Seymour and R. Thomas, A new proof of the four-colour theorem, *Electronic Research Announcements of the American Mathematical Society*, **2**: 17–25, 1996.

[44] T. L. Saaty and P. C. Kainen, *The Four-color Problem: Assaults and Conquest*, McGraw-Hill, 1977, reprinted by Dover, 1986.

[45] P. G. Tait, On the colouring of maps, *Proceedings of the Royal Society of Edinburgh*, 501–3, 1879–80.

[46] P. G. Tait, Remarks on the previous communication, *Proceedings of the Royal Society of Edinburgh*, 729, 1879–80.

[47] P. G. Tait, Note on a theorem in geometry of position, *Transactions of the Royal Society of Edinburgh*, **29**: 657–60, 1880.

[48] W. T. Tutte, How to draw a graph, *Proceedings of the London Mathematical Society*, **13**: 743–67, 1963.

[49] W. T. Tutte, A theorem on planar graphs, *Transactions of the American Mathematical Society*, **82**: 99–116, 1956.

[50] P. Wernicke, Über den kartographischen Vierfarbensatz, *Mathematische Annalen*, **58**: 413–26, 1904.

[51] A. T. White, *Graphs, Groups, and Surfaces*, North-Holland, 1973.

[52] R. J. Wilson, An Euler trail through Königsberg, *Journal of Graph Theory*, **10**: 265–75, 1986.

[53] R. J. Wilson, *Introduction to Graph Theory*, Longman, 1st ed. 1972, 2nd ed. 1979, 3rd ed. 1985.

Index

139

Printed in the United States
By Bookmasters